▶ ESSENTIAL SURVIVAL STORIES

OCEAN
SURVIVAL STORIES

BY REBECCA ROWELL

Essential Library

An Imprint of Abdo Publishing
abdobooks.com

ABDOBOOKS.COM

Published by Abdo Publishing, a division of ABDO, PO Box 398166, Minneapolis, Minnesota 55439. Copyright © 2024 by Abdo Consulting Group, Inc. International copyrights reserved in all countries. No part of this book may be reproduced in any form without written permission from the publisher. Essential Library™ is a trademark and logo of Abdo Publishing.

Printed in the United States of America, North Mankato, Minnesota.
102023
012024

Cover Photo: Eric Gevaert/Shutterstock Images
Interior Photos: Eric Gevaert/Shutterstock Images, 1; Shutterstock Images, 4–5, 12–13, 15, 16–17, 25, 26, 32, 37, 52–53, 74–75, 80, 86–87, 92; Marcel Mochet/AFP/Getty Images, 8; Red Line Editorial, 19, 40; Mr. James Kelley/Shutterstock Images, 22; Tech. Sgt. Matt Hecht/US Air National Guard/DVIDS, 29; David Wingate/Shutterstock Images, 30–31; North Yarmouth Academy/Flickr, 39; 20th Century Fox/Photofest, 43; Sheila Fitzgerald/Shutterstock Images, 44–45; Lekan Oyekanmi/AP Images, 48; Washington Imaging/Alamy, 50; STXfilms/Photofest, 55; Arne Bramsen/Alamy, 58; North Wind Picture Archives/Alamy, 62; Prisma Archivo/Alamy, 66; Chris Hellier/Alamy, 69; Mikhail Varentsov/Shutterstock Images, 71; Interim Archives/Archive Photos/Getty Images, 72; STR/AFP/Getty Images, 79; Giff Johnson/AFP/Getty Images, 84; Pat Wellenbach/AP Images, 89; Andrew Will/Shutterstock Images, 91; Dean Drobot/Shutterstock Images, 97

Editor: Marie Pearson
Series Designer: Maggie Villaume

Library of Congress Control Number: 2023939426

PUBLISHER'S CATALOGING-IN-PUBLICATION DATA

Names: Rowell, Rebecca, author.
Title: Ocean survival stories / by Rebecca Rowell
Description: Minneapolis, Minnesota: Abdo Publishing, 2024 | Series: Essential survival stories | Includes online resources and index.
Identifiers: ISBN 9781098292225 (lib. bdg.) | ISBN 9798384910169 (ebook)
Subjects: LCSH: Survival--Juvenile literature. | Adventure and adventurers--Juvenile literature. | Shipwreck survival--Juvenile literature. | Survival at sea--Juvenile literature. | Survival swimming--Juvenile literature. | Wilderness survival--Juvenile literature.
Classification: DDC 613.69--dc23

CONTENTS

CHAPTER ONE
STEVEN CALLAHAN, SEA LOVER 4

CHAPTER TWO
THE DANGERS OF THE OCEAN 16

CHAPTER THREE
STEVEN CALLAHAN, SURVIVOR 30

CHAPTER FOUR
TRAPPED UNDERWATER 44

CHAPTER FIVE
ADRIFT, HEARTBROKEN, AND ALONE 52

CHAPTER SIX
A POLAR EXPEDITION GONE WRONG 62

CHAPTER SEVEN
A RECORD-BREAKING JOURNEY 74

CHAPTER EIGHT
SURVIVING OCEANS 86

ESSENTIAL FACTS	100
GLOSSARY	102
ADDITIONAL RESOURCES	104
SOURCE NOTES	106
INDEX	110
ABOUT THE AUTHOR	112

The account in Chapter Seven mentions thoughts of suicide.

CHAPTER 1

STEVEN CALLAHAN, SEA LOVER

American Steven Callahan had loved boats for as long as he could remember. They had always been part of his life. Steven began sailing when he was 12 years old. He enjoyed being connected to the water and cherished the experience of living simply. He also enjoyed the beauty of the ocean.

Not long after learning to sail, Steven read the book *Tinkerbelle*. In it, Robert Manry told the story of his solo journey across the Atlantic Ocean in 1965. Manry and his 13.5-foot (4.1 m) boat, the *Tinkerbelle*, made the trip from the United States to England in 78 days.[1] Manry's adventure appealed to young Steven. Sailing across the Atlantic became his dream.

◀ With just a single mast, sloops are easier for one or two sailors to manage than boats with multiple masts.

BRINGING HIS DREAM TO LIFE

As he grew older, Callahan honed his sailing skills. By 16, he was sailing day trips on his own. He also sailed with others, sometimes traveling hundreds of miles along the coast. In the mid-1970s, Callahan was in his early twenties when he completed his first solo trip to another country, traveling from the United States to Bermuda.

Callahan also learned to build boats. As a teenager, he helped build one that was 40 feet (12 m) long.[2] He read more stories about real-life sailing adventures, never letting go of his dream to sail across the Atlantic.

Callahan also became a naval architect, designing and building boats. In 1980, Callahan was in his late twenties when he sold a boat and put his earnings toward building a boat for himself, which he named *Napoleon Solo*. The sloop measured a bit more than 21 feet

SEAFARING SCOUTS

Steven Callahan began sailing when he was a member of the Sea Scouts, a program of the Boy Scouts of America. His first sailing trip was with Arthur Adams, his scoutmaster. Callahan described that experience: "The first time I went [sailing] with him, everything felt right."[3]

More than 100 years old, Sea Scouts is open to people ages 14 to 20. Groups are organized across the United States. Sea Scouting takes place on the ocean and on other bodies of water, such as lakes and rivers.

(6.4 m) long.[4] Callahan loved it, writing, "*Solo* became much more than a boat to me. I knew her every nail and screw, every grain of wood. It was as if I'd created a living being."[5]

As beautiful as the *Napoleon Solo* was to Callahan, it also had to function properly and be watertight. Callahan and his friend Chris Latchem tested the new watercraft, sailing from Maryland to Massachusetts in "a harsh thousand-mile [1,600 km] shakedown . . . through late-fall gales."[6] The *Napoleon Solo* passed the test. Callahan was ready to pursue his childhood dream of sailing across the Atlantic. Now, though, it had grown to be something more than the adventure he first imagined as a boy. Sailing successfully across the Atlantic would test Callahan's skills as a sailor and a boat designer.

ROBERT MANRY AND *TINKERBELLE*

Robert Manry lived in Willowick, Ohio, and worked for a newspaper. He became interested in sailing after hearing a talk at his high school by a man who shared stories of being at sea. Manry then began reading real-life ocean stories. In 1958, Manry bought the 13.5-foot (4.1 m) *Tinkerbelle*. The boat was 30 years old. He practiced sailing on Lake Erie. On June 1, 1965, Manry departed Falmouth, Massachusetts. He reached Falmouth, England, on August 17. At the time of Manry's 78-day crossing, the *Tinkerbelle* was the smallest boat ever to have crossed the Atlantic.[7]

⚠️ The first Mini Transat race was in 1977. The race is held in odd years. Approximately 90 participants were expected for the 2023 race.

THE PLAN

In the spring of 1981, Callahan developed his plan. Later that year, he would start his journey, sailing from Rhode Island to Bermuda on his own and meeting Latchem. From there, the two men would sail to England together, and then Callahan would sail by himself in the Mini Transat, a race that would take him from Penzance, England, to Antigua, an island in the eastern Caribbean. The following spring, in early 1982, he would sail back to the United States, landing in New England.

In June 1981, Callahan set sail from Newport, Rhode Island. He had almost everything he owned with him. He reached the Bahamas, where Latchem joined him. The two men sailed to Penzance, England. Callahan had successfully completed the second leg of the planned route. He described this experience, recognizing that while he had achieved his dream, he would soon face a far greater challenge:

> *The Atlantic crossing to England with Chris was exhilarating—gales, fast runs, whales, dolphins. It was the stuff adventure is made of. And as we approached the coast of England, I felt I was ending the whole experience that had begun at my birth, and beginning a new one.*[8]

A CHANGE IN PLANS

The Mini Transat started on September 26, 1981, even though high winds made conditions troublesome. All the participants struggled, and five boats ended up sinking. Callahan quit the race before finishing, landing in La Coruña, Spain, to do some repair work on the *Napoleon Solo*.

From Spain, Callahan sailed to the Canary Islands, a group of islands near the coast of northwestern Africa. On January 28, 1982, while on the islands, Callahan sent a letter to his family telling them his plans. He would sail for Antigua, his last stop before heading home to the United States. He expected to reach Antigua around February 24. Callahan set sail for Antigua on the evening of January 28 with a three-month supply of food and water.

The first week of this leg of the trip went well. The conditions were perfect for sailing. Callahan took advantage of the situation to do things he enjoyed. He drew, read, wrote, and took photos. But things took a turn for the

> **I wish I could describe the feeling of being at sea, the anguish, frustration, and fear, the beauty that accompanies threatening spectacles, the spiritual communion with creatures in whose domain I sail.[9]**
>
> —Steven Callahan, written in Bermuda in 1981

worse on February 4. The wind changed direction and intensified. Then, in the middle of the night, something hit his boat. He thought it was perhaps a whale or a shark. The *Napoleon Solo* began filling with water. Soon the front of the boat and companionway, which is the steps leading to the lower part of the boat, were underwater.

Callahan, who had been sleeping belowdecks, jumped into action. He ran to the deck with a knife in hand and cut the line that held the life raft in place. He pulled loose the piece of the raft that made it inflate and then promptly jumped in, firmly holding the knife in his teeth.

The *Napoleon Solo* continued to stay afloat, but waves as high as nine feet (3 m) crashed over it.[10] With part of the boat still above water, Callahan collected what he could to survive on his raft. Diving underwater repeatedly to reach

NO INSURANCE NEEDED

When Steven Callahan set sail from Rhode Island in 1981, he did not have boat insurance. People purchase different types of property insurance to protect against financial loss. But insurance brokers wanted to charge high rates for the risky journey. Callahan figured the cost of materials to build a new boat would be less expensive than the cost of insurance, so he went without coverage. He thought the worst thing that could happen on his trip was dying, and he would not care about money if he were dead. The second-worst thing would be losing his boat, but he knew he would ultimately be okay if that happened.

▲ Callahan departed from the Canary Island called El Hierro, which is 19 miles (30 km) long.

the cabin, he managed to get some essentials, including food, a sleeping bag, a speargun, water, and flares. A flare is used to attract the attention of anyone within sight. He piled the supplies into his raft.

Ideally, Callahan would plug the holes in the *Napoleon Solo* and stop the water from coming in. After that, he could bail the water that was already flooding the craft. But the

seas got rougher. Then, as sunrise approached, the 30-foot (9 m) rope connecting the raft to the *Napoleon Solo* broke. Callahan was alone on a nylon raft 800 miles (1,290 km) from his destination and 450 miles (724 km) from the closest shipping routes between New York and South Africa, where he might attract someone's attention and be rescued.[11] His water supply was perhaps enough to survive for two weeks, but he had almost no food. He did have three saltwater stills, devices that remove salt from seawater and make it drinkable.

But no one knew Callahan was in trouble, and his family would not have a clue for about three weeks, which was when he was due to arrive in Antigua. By then, he might be dead. With few provisions, no shelter, no means of communication, sharks and other sea creatures below, and no land or help in sight, Callahan's chances of surviving were not good.

OCEAN SURVIVAL

Every day, people take to the ocean. Some do it for work, such as for shipping companies or as members of the military. Some do it for fun. They sail alone, with loved ones, or with hundreds of people aboard a cruise ship. Still others do it as a means of escape, taking to the water in search of a better life in a new country.

Whatever the reason, boating on the ocean does not always go as planned. Every year, people around the world experience trouble at sea. The fortunate ones live to share their stories, many of them reaching safety with the help of others. The ocean environment, while beautiful and awe

▲ Today, forecasting can help people avoid getting caught in a storm on short trips. But when sailing the open ocean, it may be impossible to avoid a storm that builds up several days into a trip.

inspiring, can present immense challenges even for people with much experience in its waters. But some people face those challenges and survive. From their stories, others can learn how to prepare for the worst and increase their own chances of preventing or surviving a disaster at sea.

CHAPTER 2

THE DANGERS OF
THE OCEAN

Earth is sometimes referred to as the Blue Planet because the ocean covers more than 70 percent of its surface, making it mostly blue when viewed from space.[1] The ocean is spread throughout the world. This singular body of water is called the world ocean, and people know smaller parts of it by different names. The five major divisions are the Arctic, Atlantic, Indian, Pacific, and Southern (Antarctic) Oceans.

Ocean water is not consistent worldwide. Characteristics such as temperature, salinity, and currents vary by location, depth, and season. Because the ocean stretches from frozen areas to tropical ones, its temperature varies widely by region. It can range

◀ Ocean waves can be destructive and deadly. Wind causes the surface of the water to rise and fall in waves.

from less than 32 degrees Fahrenheit (0°C) to higher than 86 degrees Fahrenheit (30°C).[2]

Ocean temperature also varies from the surface to the seafloor. In most parts of the ocean, the temperature drops the farther down one goes. For example, in the tropics the sun keeps the surface water warm year-round. But in the thermocline, which ranges from 660 to 3,300 feet (200–1,000 m) below the surface, the temperature drops.[3] There, it ranges from 46 to 50 degrees Fahrenheit (8–10°C).[4]

However, the depths aren't always cold. Hydrothermal vents are areas with hot springs where seawater enters Earth's crust, gets heated, and then returns to the ocean. This warms the surrounding water along the ocean floor.

Another variable is salinity, which is the amount of salt that seawater has. In surface

THE OCEAN IS CLEAR, NOT BLUE

While the ocean often appears to be blue, the water is actually clear. The appearance of color is the result of different factors. One is that the water acts as a mirror and reflects the sky. The bigger reason is that particles in the water bounce back the blue light in sunlight traveling through the water. But the ocean does not always look blue. Rough waters can appear white because of air bubbles, and rain gets in the way of light bouncing off the water and can cause the ocean to appear a darker grayish green. Sometimes, the ocean has areas of bright color from living things such as algae, which can make the water a bright green.

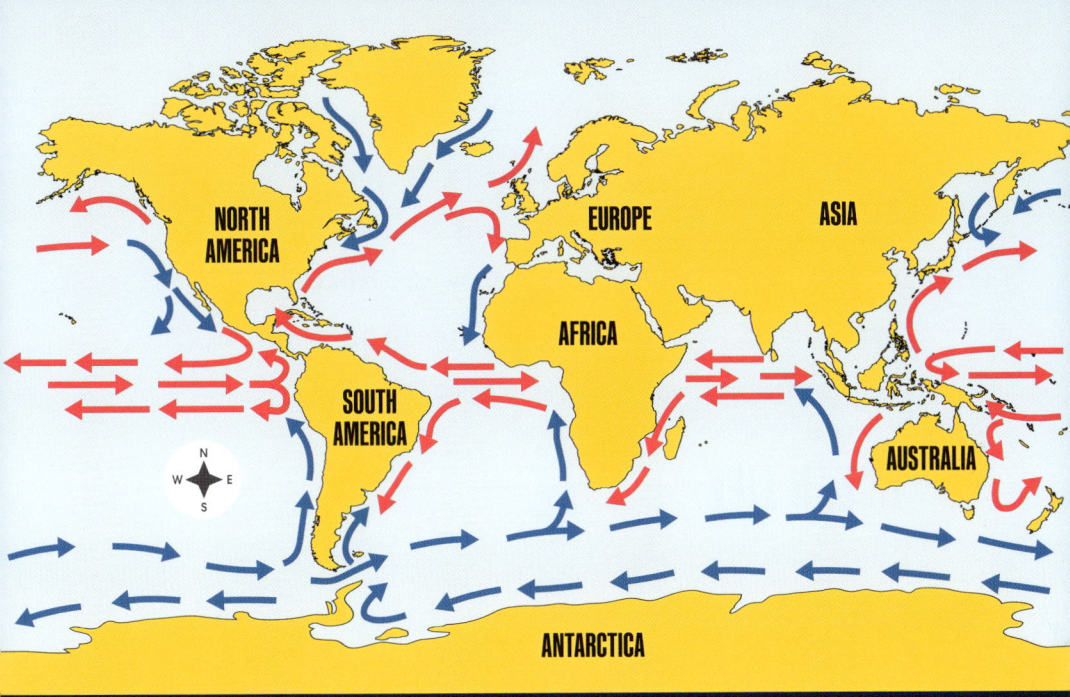

▲ Some maps show the direction of ocean currents as well as whether they are warm currents, *red*, or cool, *blue*.

waters, rainfall adds water, which decreases salt content, while the formation of ice removes water and increases salt content. Salinity affects the temperature at which seawater freezes. The more salt the water has, the lower its freezing temperature will be. That means the temperature could be 32 degrees Fahrenheit (0°C) and the water will not freeze. Because rainfall and the formation of ice affect only the waters nearer to the surface, salinity at ocean depths is almost constant.

A consistent feature of seawater worldwide is that it does not stop moving. It has different currents at different depths. The ocean's surface also has numerous currents. These areas move with the help of wind. Currents can affect where a boat with a dead engine or someone who falls overboard might drift. Currents also affect where fish go, which can benefit people trying to bring in a large catch.

OCEAN LIFE

The ocean is the planet's biggest ecosystem and is home to a variety of life. Thousands of plants grow in seawater. In shallower areas near shorelines, beds of seagrass can grow in warm water, while kelp forests thrive in cold water. Both plants provide food and shelter for other sea life. For example, manatees eat seagrass. Algae and fungi also grow in the ocean.

Like seagrass beds and kelp forests, coral reefs are located near coastlines and in warm and cold waters. Corals exist in shallow and deep waters. Some types of corals look like plants, and others have a rough appearance and can seem like rock. Corals are composed of tiny living animals called polyps. They cluster together in colonies. Each polyp releases calcium carbonate to create an external skeleton, and the skeletons form the structure of coral. Some corals have the longest life spans of all animals on Earth, living up to 5,000 years.[5]

Scientists estimate that more than one million animal species may inhabit the ocean, including corals.[6] Ocean animals live in all depths, from the surface to the ocean floor, and in all temperatures. Jellyfish and worms move with ocean currents, while mollusks, crustaceans, and sea stars live on the seabed.

Approximately 20,000 species of fish inhabit the ocean.[7] Reptiles such as turtles and crocodiles live there too. Ocean birds include ducks, pelicans, and penguins. And mammals such as whales, dolphins, and seals swim in the world's seas.

The ocean can be gorgeous and amazing. But being lost at sea can make the ocean challenging, scary, and deadly. Several factors play a role in the severity of the experience, including sun exposure, the water temperature, the presence

GREAT BARRIER REEF

The Great Barrier Reef is a famous natural landmark made of coral. It is not a single reef but a system of reefs that includes approximately 3,000 coral reefs. Visible from outer space, the Great Barrier Reef covers 134,600 square miles (348,700 sq km) and is more than 1,430 miles (2,300 km) long. The reef is home to an assortment of wildlife, including 17 species of snakes; at least 330 species of ascidians, which are invertebrates sometimes called sea squirts; and more than 1,500 species of fish. And many more animals visit the area but do not live there all the time, including six species of sea turtles; 30 species of whales, dolphins, and porpoises; and 215 species of birds.[8]

▲ Coral reefs provide hiding places for prey and predators alike, including moray eels.

of sharks, whether the person is in a boat, and the amount of drinkable water available.

DEHYDRATION

Humans require water to survive. Water keeps the body operating the way it should. A person loses water in different ways, including through breathing, feces, sweat, and urine. Being without fresh water can quickly lead to dehydration, which occurs when the body releases more water than it takes in. For example, in a hot environment, a person can begin to experience dehydration in as little as 60 minutes.[9]

Dehydration has a variety of symptoms. Someone suffering from dehydration may be thirsty, have a dry mouth, feel tired, or feel lightheaded. The individual may also urinate less frequently than usual, and their urine may be dark and have a strong smell. Being dehydrated can have serious consequences. A lack of water and the dehydration that results can lead to seizures, brain damage, and even death.

Although the ocean is water, humans should not drink it because of the salt. Besides simply not tasting good, the amount of salt in seawater is too much for the body to handle. The kidneys remove impurities from the body's blood, including salt. Because water from the ocean has so much salt, the kidneys would need more water than a person would get from the seawater to remove it. So the kidneys would take water from the body to process the salt. That would lead to dehydration or speed up the dehydration already in progress. Even in ideal weather conditions, such as moderate temperatures, a person adrift on the ocean with no water would last less than one week.

SUNBURN

Being stranded without shelter can result in multiple issues. One is sunburn. Severe cases of sunburn can cause second-degree burns. This level of burn goes deeper than the top layer of skin. Second-degree burns cause blisters, and sometimes skin will get thicker. The skin may also turn red and hurt.

The pain from these burns can be intense. Different treatments for the pain and the wound include running cool water over the burned area, taking painkillers, and putting antibiotic cream on the blisters. But these options are often not available to someone stranded at sea.

HEAT EXHAUSTION AND HEATSTROKE

For those adrift in tropical waters, a lack of shelter can also lead to heat-related illnesses, such as heat exhaustion and heatstroke. Heat exhaustion occurs when the body has lost a lot of water and salt, usually by sweating heavily. Symptoms include dizziness, headache, nausea, thirst, weakness, increased body temperature, and decreased urination.

▲ Depending on the craft, sometimes the only shade a person on the ocean has is what they bring with them.

In addition, someone suffering from heat exhaustion may be crabby and impatient.

Heatstroke is worse than heat exhaustion and is the most dangerous heat-related illness. When someone is suffering from heatstroke, their body is unable to regulate its temperature as usual. As a result, the body's temperature rises quickly, the sweating function stops working properly, and the body cannot cool down. In just 10 to 15 minutes, the body's temperature can rise to a dangerous 106 degrees Fahrenheit (41°C) or higher.[10] The normal temperature for the human body is approximately 98.6 degrees Fahrenheit (37°C).[11]

Heatstroke, like heat exhaustion, has physical and mental symptoms. These include confusion, slurred speech, extremely high body temperature, seizures, and skin that feels hot and dry or sweats excessively. The person may pass out and slip into a coma.

Providing shade and applying cool water to the skin are treatments for heat exhaustion and heatstroke. Medical care is also important, especially for those suffering from heatstroke. Without it, the sufferer may die or become permanently disabled.

HYPOTHERMIA

Hypothermia is another issue a person at sea might face. Hypothermia is when the body gets too cold. It happens when the body loses heat more quickly than it can make heat. Hypothermia occurs when body temperature is less than 95 degrees Fahrenheit (35°C).[12]

▼ People should be aware of both daytime and nighttime temperatures in the area of ocean where they will be traveling and bring appropriate clothes.

Similar to how the body begins to falter when it becomes too hot, it also malfunctions when it is too cold. The nervous system is unable to work properly. Hypothermia has a variety of symptoms. They include shivering, mumbling or slurred speech, decreased breathing, weak pulse, clumsiness, extremely low energy or drowsiness, confusion or memory loss, and unconsciousness. At the severe stages, organs will also stop working properly. Ultimately, hypothermia can cause death by interfering with the heart and respiratory system.

Most ocean water is cold, so getting wet can result in hypothermia. That makes staying dry important. Piling on blankets and other items can help keep a person warm. For people stranded together, hugging can help them warm up, and getting under a blanket or into a sleeping bag, if one is available, can increase warmth. It's important to remember to cover as much of the body as possible, including the head, to prevent heat loss.

SHARKS

The ocean itself poses a danger to humans because of some of the creatures in it, including sharks. This danger is increased for people stranded in the water instead of on

SHARK ATTACKS IN 2022

The University of Florida's Florida Museum of Natural History tracks shark attacks worldwide. In 2022, the museum examined 108 reported events between sharks and humans and recorded 57 unprovoked bites across seven countries. Unprovoked bites are when humans are in the shark's natural habitat but have not done anything to interact with the shark. The United States had the highest number with 41. Most of those occurred in Florida, which had 16. The United States had one fatality. The other countries with unprovoked shark attacks were Australia (9), Egypt (2), South Africa (2), Brazil (1), New Zealand (1), and Thailand (1).[13]

a vessel. It is important that people stay in their vessel if possible to avoid the risk of being bitten by a shark. But even those in a boat or raft are not fully protected from a shark attack. Sharks sometimes bite boats, including rubber ones. This poses a danger because a shark bite could puncture the vessel's hull and potentially sink it.

People faced with a shark attack can try different techniques to save themselves. One option is to punch the creature in the nose. Other options are to pull the shark's gills or poke it in the eye, making sure to keep hands away from its mouth.

DROWNING

The ocean is also dangerous because it poses a risk of drowning. A wave might overturn a boat or wash a person

▲ The US Coast Guard has many roles, including conducting search and rescue operations along the United States' ocean coasts and interior bodies of water.

overboard, or someone may simply slip and fall off the vessel. Even a person who knows how to swim could drown. The chances of drowning are greater than of being attacked by a shark.

In the United States, hundreds of people die each year from drowning while sailing. According to the US Coast Guard, the number of deaths associated with recreational boating in any body of water was 636 in 2022. Of those, 75 percent were the result of drowning.[14]

Whatever the circumstances, being stranded in the ocean presents numerous challenges and risks. The ocean can be deadly. Still, even in the most dire of situations, people have survived.

CHAPTER 3

STEVEN CALLAHAN,
SURVIVOR

Steven Callahan was alone on a raft in the middle of the Atlantic Ocean. With the sinking of the *Napoleon Solo*, his new vessel was a six-foot (1.8 m) round raft he dubbed *Rubber Ducky III*.[1] It had a canopy that could provide some protection. But it could not keep out water, and that first night in February 1982, Callahan used a can to bail out water when waves splashed into the raft.

He had managed to retrieve three pounds (1.4 kg) of food and four quarts (3.8 L) of water from the *Napoleon Solo*.[2] Assuming the raft stayed afloat, Callahan thought he could survive for 15 days with those supplies.

◀ Some life rafts have covers, but holes for the passengers to enter and exit allow water to get inside.

Callahan had also retrieved tools from the *Napoleon Solo* that further increased his chances of surviving. He had a speargun he bought in the Canary Islands on a whim that he could use to fish. He also had three solar stills, but he got only two to work. Together, they created approximately 1.3 quarts (1.2 L) of drinkable water a day.[3] Having a working still was necessary, but the amount of water Callahan's two stills produced was well below the 3.9 quarts (3.7 L) recommended daily for an adult man.[4]

The seven flares Callahan grabbed from the *Napoleon Solo* could save his life. When he saw a ship approximately 14 days after the *Napoleon Solo* sank, Callahan shot a flare

▼ There are many types of emergency flares. Some are launched into the air, while others are held by hand or floated on the water.

into the air to catch the attention of someone aboard. The ship appeared to get closer, so he sent five more into the air, taking several gulps of water as he waited. But the ship kept on its path, and Callahan scolded himself for using so many flares and drinking so much water. He saw a few more ships, but no one aboard saw him adrift. Each vessel continued on its journey.

PHYSICAL DISCOMFORT

Callahan suffered physically. From the outset, he was in pain. Sailing the Atlantic continually exposed him to salt water, which is hard on human skin. His clothes became soaked and stayed that way. As a result, his skin developed dozens of boils. Callahan kept a daily log while floating on the Atlantic. On day two of being adrift, he wrote about his experience with the boils:

> *They multiply quickly under my wet T-shirt and sleeping bag. Gouges and abrasions cover my lower spine, butt, and knees. They are foul, but I suppose they are clean. I'm often awakened with the searing pain of salt burning their putrid tenderness. The raft is too small for me to stretch out in, so I must rest curled up on my side. At least this helps to keep the cuts dry.[5]*

Hunger and thirst also challenged Callahan. He quickly found himself longing for food and water. On February 8, day four, he wrote in his log, "My mind creates fantasies of food and drink and turns continually back to *Solo*, to the pounds of fruits, nuts, and vegetables and the gallons of water within her."[6]

Callahan rationed his drinking water, knowing the limits of the stills. He managed to catch fish, but his food intake was limited, and he was feeling the effects of starvation. By day 13, his body had used its store of fat. Callahan noted, "The fat is gone. Now my muscles feed on themselves."[7] As a result, his body moved more slowly and he tired more quickly.

STAYING STRONG

Alone and adrift, with no sense of when or even if

CALLAHAN'S FAMILY

While Callahan waited, so did his family. Callahan's father contacted the US Coast Guard on March 9, 1982, about Callahan being missing. For two weeks, the Coast Guard sent out an alert about the *Napoleon Solo* being missing, but no one reported seeing the vessel. Callahan's brother, Ed, flew to their parents' house and made finding Callahan his full-time job. Ed, their father, and their friends tried to estimate where Callahan might be and reached out to the Coast Guard again, but there was no sign of him. By mid-April, even Callahan's sailing friends figured he was gone. A week later, Callahan's family got the good news that he had been found and was alive.

rescue might occur, staying hopeful was challenging. Callahan was a knowledgeable and experienced sailor, but this situation tested him.

Being adrift at sea gave Callahan time to think. He found himself pondering his life and feeling bad. He questioned personal choices and even his ability as a sailor. He wanted to survive so he could do better moving forward. Callahan said, "I regretted every mistake I'd ever made—I was divorced, and felt I had failed at human relations generally, at business and now even at sailing. I desperately wanted to get through it so I could make a better job of my life."[8]

Fear was also constant. He worried that he would not be able to catch food or have enough drinking water. He was concerned that a shark or other sea creature might damage his raft. He fretted that his body would become too weak for him to function or that his mind would give in to the fear.

An especially difficult part of his time in the *Rubber Ducky III* came when the bottom tube of the raft became damaged. It was the morning of day 43, and several mahi-mahi, a type of fish also known as dolphinfish, flanked the raft. Callahan speared one, and as it fought the line, it bumped the raft, sending the tip of the spear into the raft's side. The top tube remained inflated, but water poured in.

For several days, Callahan tried to fix the hole. He covered it with a patch, but it came off every time he added air.

While dealing with the hole, the weather turned bad. Fixing the hole required putting his hands in the water. Sometimes, sharks approached. The situation got the better of Callahan. He yelled and cried in frustration.

Callahan felt ready to give up. But he had to keep trying to fix the hole if he wanted to survive. So he told himself to keep thinking about the problem. He took another look at his supplies, trying to figure out what could be useful. That is when he figured out a solution.

> **The sea teaches you to make do with what you've got, plus what surrounds you in the environment.**[10]
>
> —Steven Callahan in an interview on January 31, 1986

Among Callahan's pieces of equipment was a Boy Scout utensil kit, which included a fork, a knife, and a spoon made from stainless steel. Callahan was able to get the fork to hold the torn nylon together. It was one week after the hole first appeared. He filled the bottom tube with air, and it held. He described the experience: "I got scared by the thought I would be dead in a few hours; I found a way to fix the raft, and it felt like the biggest victory of my life."[9]

▲ The mahi-mahi can grow up to six feet (1.8 m) long. It can swim more than 55 miles per hour (90 kmh).

WAITING

Callahan had resolved the raft problem, but he was still under considerable stress. It was day 50, and his only hope of survival was either drifting to land or being found by someone. It was a waiting game, and Callahan's goal was simply to stay alive.

Being adrift was taking a toll on Callahan. He said in an interview, "My body and mind were shutting down; it was as if I could feel all the people who had ever been lost at sea around me. I had no more to give."[11]

But Callahan did not give up. He fought the challenges of the environment and his mind. On April 2, day 57, he created a makeshift sextant, a tool used to navigate by measuring the angle between the horizon and the sun or a star. With that angle measurement, the time of day, and a navigation chart as reference, people can figure out their latitude and longitude. Callahan's sextant was three pencils rigged together.

That day, Callahan determined his latitude and longitude. He used these rough measurements to adjust his course. But the makeshift sextant was not guaranteed to be accurate. Being off even a small amount could add weeks to his dangerous journey.

On April 6, day 61, the ocean changed. Callahan drifted from clear waters into a massive patch of seaweed. Knowing the variety of sea life that could be among it, he quickly pulled some of

THE MAHI-MAHI

During his months drifting in the Atlantic Ocean, Callahan encountered various sea creatures. The mahi-mahi became important to him. These fish often surrounded his raft, and he sometimes managed to spear one. That made them an important source of food. But they were more to Callahan. "The fish offered me real companionship," he said. "They nourished me. They almost killed me. And in the end, they brought me salvation."[12]

the seaweed aboard. The ocean rewarded him with shrimp and crabs.

Two days later brought a different excitement. On April 8, Callahan saw a ship. It was the seventh he had seen during his 63 days adrift. He sent up a flare. It was daytime, so the flare was not as noticeable as it would be at night. The ship continued on its route. Callahan decided reaching land would be his only hope. He was able to spear a mahi-mahi, which would give him food for a few days.

On April 10, day 65, he grabbed a bird that landed on the canopy. It provided another meal. Even with this fortune, continuing was difficult. Callahan described himself on April 12, day 67, writing in his log, "I feel swayed more

▼ Callahan travels to schools and many other places to speak about his experiences and survival strategies.

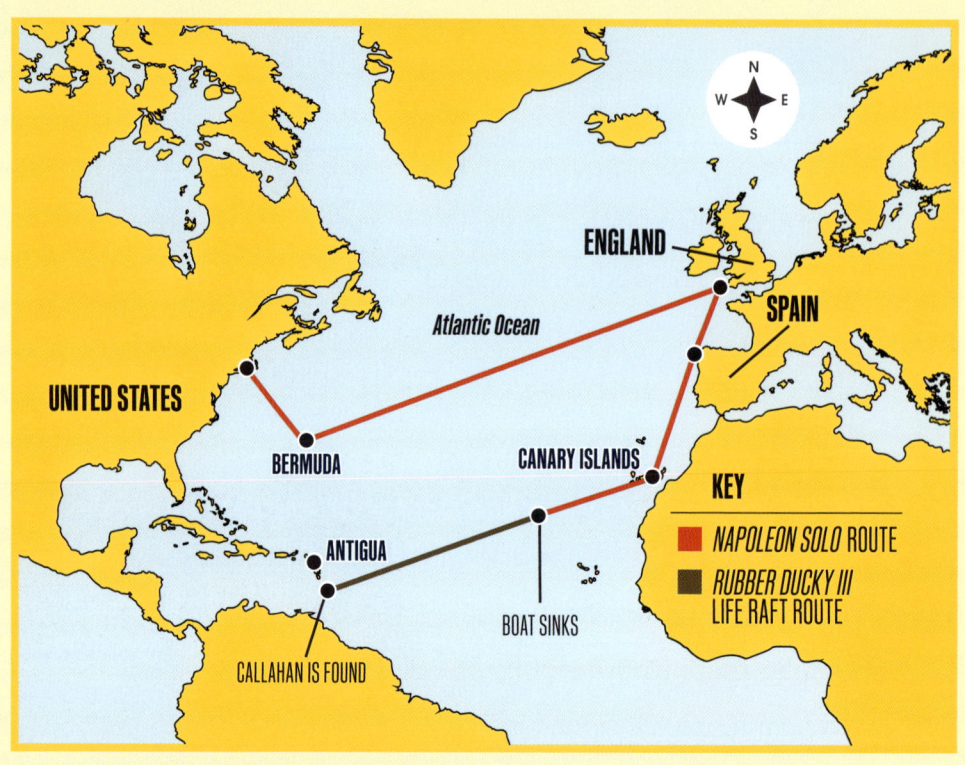

▲ When his boat, the *Napoleon Solo*, capsized, Callahan's trip home across the Atlantic Ocean took a new course.

and more by my body's demands, feel stretched so tight between my body, mind, and spirit that I might snap at any moment."[13] On day 71, his solar stills stopped working. Without them, he could not make the seawater drinkable. He had only three cans of water left, and they would be gone quickly. He was sometimes able to capture rainwater for drinking, but he needed more water than was available.

FINALLY

Day 75 brought hope. That evening of April 20, Callahan spotted land. The following morning, he focused on steering his raft toward it. At the same time, three men fishing near Guadeloupe's Marie Galante Island noticed birds focused on something in the water. It was the *Rubber Ducky III*. The birds were interested in fish guts Callahan had thrown back into the ocean after catching and eating a mahi-mahi.

THE STORY OF POON LIM

Poon Lim was part of the crew of the SS *Benlomond*, a British ship. In November 1942, during World War II (1939–1945), the ship was traveling from South Africa to Suriname in South America. During the voyage, a German submarine hit the ship with a torpedo. Lim jumped into the sea and was the only survivor. He found a wooden raft from the *Benlomond* in the water. It had some food and water. When these supplies were gone, Lim survived by collecting rainwater and by fishing. On April 5, 1943, three fishermen rescued Lim off the coast of Brazil. He had been adrift for 133 days. Lim later traveled the world to share his story and teach people how to survive if lost at sea.

The three men usually fished the other side of the island, but not that day. They saw the raft and headed to it. On April 21, 1982, the men rescued Callahan. He had been adrift in the *Rubber Ducky III* for 76 days, floating from near West Africa to the Caribbean Sea.

After his rescue, Callahan needed time to recover. It took six weeks for him to walk normally again.

Callahan's knowledge and experience helped keep him alive. He also realized his good fortune. In an interview about his rescue, Callahan shared the details of his ordeal. Summing up the experience, Callahan said:

> It's . . . so awesome and so incomprehensibly complicated and incredible that I do not ever expect to even understand it. For example, in the raft voyage, several things happened which, if not miraculous, certainly border on the miraculous. Many things happened to me that I could so easily abstract meaning from—coincidences which saved me, coincidences which caused me trouble. So many things fit together that to me it is a great mystery.[14]

Callahan developed his ocean log into the book *Adrift: 76 Days Lost at Sea*, which was first published in 1986. It was on the *New York Times* bestseller list for more than 36 weeks.[15] It has been translated into 15 languages.[16]

Callahan discovered that many readers can relate to his story. He noted, "The details of their particular survival stories may vary, but the point is they've gone through it. And, you know, they really do surprise themselves with their ability to endure."[17]

CULTURAL IMPACT
CALLAHAN AND *LIFE OF PI*

In 2012, *Life of Pi* hit movie theaters. Based on the 2001 novel of the same name by Yann Martel, the film told the story of a boy stuck at sea after a shipwreck. Suraj Sharma was the actor playing Pi, the story's main character. To help Sharma portray his situation as authentically as possible, director Ang Lee hired someone who had survived a similarly harrowing experience: Steven Callahan. Callahan knew the fear, stress, loneliness, desperation, and hopelessness that being adrift on the ocean can cause. He worked with Sharma to help him express the reality of being deserted at sea.

In early 2013, *Life of Pi* received 11 Oscar nominations and won four of the 11 categories in which it was nominated.[18] In February of that year, before the awards ceremony was held, National Public Radio interviewed Callahan. He spoke about his experience of being adrift for 76 days. He also discussed working on *Life of Pi*, sharing, "For me, it became quite the personal voyage because it's a continuum of my own experience 30 years ago. And to me, it's kind of amazing that the ocean that kept me alive would allow me to go down this path . . . to make the ocean, as [Ang] said, a major character."[19]

▲ Pi needs to build a life raft for himself in the film.

CHAPTER 4

TRAPPED UNDERWATER

May 26, 2013, started like any other workday for Harrison Okene. The 29-year-old Nigerian was a cook on the *Jascon-4*, a tugboat. Though relatively small, tugboats are sturdy and strong, built to push and pull larger vessels. Okene and the rest of the crew were experienced seafarers. That day, they were in the Atlantic Ocean, approximately 19 miles (30 km) off the coast of Nigeria.[1] The *Jascon-4* was helping tow a tanker, a large ship filled with oil.

It was about five o'clock in the morning, and Okene had just gotten up. He was in the bathroom when everything changed. A massive wave hit the *Jascon-4* so hard the boat capsized. Okene ran out of

◀ Tugboats can help move ships in many ways, including by pushing them. They can also use towlines to pull ships.

45

CHAPTER 5

TRAPPED UNDERWATER

May 26, 2013, started like any other workday for Harrison Okene. The 29-year-old Nigerian was a cook on the *Jascon-4*, a tugboat. Though relatively small, tugboats are sturdy and strong, built to push and pull larger vessels. Okene and the rest of the crew were experienced seafarers. That day, they were in the Atlantic Ocean, approximately 19 miles (30 km) off the coast of Nigeria.[1] The *Jascon-4* was helping tow a tanker, a large ship filled with oil.

It was about five o'clock in the morning, and Okene had just gotten up. He was in the bathroom when everything changed. A massive wave hit the *Jascon-4* so hard the boat capsized. Okene ran out of

◀ Tugboats can help move ships in many ways, including by pushing them. They can also use towlines to pull ships.

45

his cabin. He had no time to dress and was wearing only his underwear. Water was filling the tugboat. The *Jascon-4* was sinking. He needed to escape.

Okene could not get to the emergency exit. The force of the water entering the ship moved him along. He ended up in the cabin of the ship's officer, where water pushed him into the bathroom. The *Jascon-4* sank, falling through the Atlantic's cold, dark waters and settling approximately 98 feet (30 m) below the water's surface.[2]

TRAPPED

As the *Jascon-4* sank, Okene ended up in a pocket of air that was trapped in the bathroom. It measured 477 cubic feet (13.5 cubic m).[3] The oxygen allowed him to continue breathing and would keep him alive for a while, but conditions were far from ideal.

TONY BULLIMORE

In 1997, Tony Bullimore was taking part in the Vendée Globe, an around-the-world yacht race sailed solo, when he encountered a storm in the Southern Ocean. The British man's boat capsized in frigid waters near Antarctica. Fortunately for 57-year-old Bullimore, he was able to send a distress signal. He was also in a wet suit, which kept him from experiencing hypothermia.

Bullimore stayed under his overturned craft, where ocean water sloshed in and out a broken window. Eventually, the Australian navy found Bullimore and several competitors who had been stranded as a result of the storm. He had spent four days adrift at sea, an experience he described as being "like a washing machine from hell."[4]

Okene was in total darkness, and he had no food or drinking water, only one bottle of Coke. Additionally, the water was frigid. The air bubble allowed Okene to keep his chest and head above water, which helped his body cope with the freezing water. He described this experience: "All around me was just black and noisy. I was crying and calling on Jesus to rescue me. I prayed so hard. I was so hungry and thirsty and cold and I was just praying to see some kind of light."[5]

> The fear alone can kill you. I took fear off me, and I believed that 'what will be, will be.' Believe in yourself and keep your faith and your mind strong.[6]
>
> —Harrison Okene, 2022, discussing being trapped underwater for 60 hours

Okene ventured from the air pocket. He used a rope he found to guide himself back to the spot after exploring in the dark. He was trying to find a way out of the sunken boat, but he did not succeed.

Time passed. Okene thought about his wife and family. He prayed and recited a passage from the Bible. He heard a noise that frightened him. He thought it was fish eating something. Because of the darkness, he could not see, and he was scared. Remaining hopeful was difficult, but then something unexpected happened.

▲ Harrison Okene, *right*, thought about his wife, Akpovona, *left*, a lot while he was trapped underwater.

RESCUED

Okene had been trapped underwater for more than two days when he heard a knocking sound. Then he saw light. It was from the flashlight of Nico van Heerden, a diver. He worked for the company DCN Diving and was tasked with finding the bodies of the crew of the *Jascon-4*. The DCN team had already recovered the bodies of a few of Okene's crewmates and expected anyone else they found to also be dead. The diver was close enough for Okene to touch. He tapped van Heerden on his arm.

Van Heerden saw Okene's hand but figured it was another dead crew member. When the diver reached for

Okene's hand, Okene grabbed him, startling him. The diver went to assist Okene and found Okene panting.

Okene was experiencing hypercapnia. This condition occurs when people are exposed to high levels of carbon dioxide. Being in the air bubble had kept Okene alive, but it had also trapped the carbon dioxide Okene exhaled when he breathed. That carbon dioxide had been accumulating for the 60 hours Okene had been in the bubble.

Symptoms of hypercapnia include rapid breathing, heart malfunction, and decreased brain function. Alex Gibbs is a DCN life support technician who was at the ocean's surface when Okene was discovered. According to Gibbs, Okene was near death.

The rescue team responded quickly when it found Okene. Van Heerden gave Okene a bottle of water to drink, and then he outfitted Okene with diving equipment. Then van Heerden guided Okene out of the sunken ship and to a diving bell, which is

DCN DIVING

The Vriens Diving Company launched on January 1, 1957, in the Netherlands. Over the years, the company and its services grew. The company became Duik Combinatie Noordersluis (DCN) on January 1, 1992. Today, DCN provides a variety of diving- and water-related services, such as building bridges and maintaining dams. Services also include recovery work such as the job that resulted in Okene's rescue.

a piece of equipment that moves divers between the ocean surface and underwater.

After reaching the diving bell, Okene was not out of danger. The ocean exerts considerable pressure that increases with depth. When a person moves from this high pressure to the lower pressure of the surface, they need to move slowly between the pressure differences to keep from getting decompression sickness. Commonly called the bends, this condition occurs when nitrogen in the blood forms bubbles. If the pressure outside the body

▼ There are two types of diving bells. The open diving bell has a dome that traps air but a bottom that is open to the water, while the closed diving bell, *pictured*, can be completely sealed and have its pressure adjusted.

drops too fast, those bubbles can burst. The situation can cause considerable pain and even death.

The diving bell moved Okene safely to the surface by maintaining a specific pressure. Once at the surface, the DCN team placed Okene in a decompression chamber. This equipment is designed to help divers' bodies decompress. In a decompression chamber, a person waits as the pressure changes gradually over a length of time. It starts with the pressure the body was at and then slowly adjusts to match the pressure outside the chamber. Doing this allows the body to release the nitrogen trapped in it without causing injury.

Gibbs looked after Okene while he decompressed. After three days of decompressing, Okene's ordeal was finally over. In the end, the divers recovered the bodies of ten of Okene's crewmates, but one person was never located.[7] Okene was the only survivor.

FROM SURVIVOR TO DIVER

Harrison Okene's experience of being trapped underwater for 60 hours did not keep him from returning to the ocean. Okene's short trip from the wreckage to the diving bell was his first experience as a diver. It was not his last. After surviving the sinking of the *Jascon-4*, Okene took up diving. He is a certified diver and can go as deep as 160 feet (50 m). Okene said of his experience as a diver, "I'm enjoying diving, it's life for me, it's fun. I believe the ocean is my world. I feel more comfortable, relaxed there."[8]

CHAPTER 5

ADRIFT, HEARTBROKEN, AND ALONE

In 1983, an American woman named Tami Oldham and a British man named Richard Sharp were in their twenties, in love, and engaged to be married. In September, they were in Tahiti and planning to return to the United States. The trip across the Pacific Ocean to Oldham's hometown of San Diego, California, was more than 4,000 miles (6,440 km).[1] They were not going to fly. They were going to sail. Sharp had gotten a job delivering the *Hazana*, a 44-foot (13 m) yacht.[2] He and Oldham would sail the Pacific together, and they planned to do it without any additional crew.

> ◀ Many yachts do not have sails, but having a sail can reduce the risk of being stranded at sea, since a sail can move the boat if the motor dies. Carrying an extra sail is helpful in case the original sail gets damaged.

Both were experienced sailors and had spent the last few months sailing around the waters of the Pacific.

SETTING SAIL

On September 22, Oldham and Sharp set sail from the Tahitian city Papeete. The couple expected to be at sea for a month. The conditions were ideal. As Oldham described them, "Great weather, great wind, and great company."[3]

During the first week, some windy days proved challenging. They worked hard managing the sails to keep the boat on course. The work was tiring, but the boat held up.

On the fifth day, seawater splashing aboard wrecked the boat's two-way communication radio. Fortunately, the couple still had a radio to receive weather information. But rough waters made sleeping difficult, rocking the boat and whipping the sails. The situation soon improved. Calmer seas for a few days gave them a break from controlling the sails.

On October 2, they crossed the equator, moving from the Southern Hemisphere to the Northern Hemisphere. That day, they also enjoyed an encounter with a pod of whales. Oldham and Sharp's journey was going well. But that would soon change.

▲ Actors Sam Claflin, *left*, and Shailene Woodley portrayed Sharp and Oldham in the 2018 film *Adrift*.

A CHANGE IN THE WEATHER

On October 8, the sunshine disappeared, replaced by gray skies and rain. And the wind seemed to blow from every direction. The next day, the weather channel WWV reported on two weather systems near Central America. Raymond was a tropical storm that was strengthening. Sonia was a tropical storm that was weakening. Even with distance between the *Hazana* and the storms, the strong winds they generated challenged the couple as they tried to manage the yacht's sails and rigging. Adjusting the main sail to shift their course took them two hours. They were exhausted, wet, and hungry. Sharp made an entry in the boat's log: "We're OK."[4]

On the morning of October 11, WWV said that Raymond had strengthened into a hurricane. The storm had also changed its path. The couple quickly got to work to change course, wanting to avoid the storm. But a weather update that afternoon brought bad news. Oldham and Sharp were heading straight for Raymond.

The couple got to work again with the rigging to change the yacht's direction. They were confident in the *Hazana*. It was designed for stormy ocean travel. But one or both of the travelers could get hurt. Sharp wrote in his log, "All we can do is pray."[5]

A TURN FOR THE WORSE

The morning of October 12 provided a bit of sunlight, but the wind and rough waters worsened. In a matter of hours, the wind speed increased dramatically. By noon, it was 115 miles per hour (185 kmh).[6]

Sharp gave Oldham their emergency position-indicating device. It was the only one they had, and it could help rescuers locate them. He went belowdecks to get another weather update and heard only static.

Taking the radio to the deck could possibly help with reception, but then he might lose it because of the water that kept coming over the boat's edge. The wind howled, the water raged, and the yacht shot high and dove low in the rough seas. The wind blew 160 miles per hour (260 kmh), causing waves 40 feet (12 m) high.[7]

Oldham asked if this was it for them. Sharp said it was not: "No. Hang on, love. . . . Someday we'll tell our grandchildren how we survived Hurricane Raymond."[8] He insisted they would survive and sent Oldham belowdecks. She headed to the cabin, leaving Sharp in the boat's cockpit, where he had secured himself to the *Hazana* with the tether of his safety harness.

At one o'clock in the afternoon, Oldham was shut in the cabin. She had connected the tether of her safety harness to a table and was in a hammock, gripping a blanket close and swinging sharply. She closed her eyes in fear. Suddenly, the yacht settled and was eerily quiet. Oldham sensed that something was wrong and felt extremely afraid. Sharp, who

HURRICANE RAYMOND SURVIVORS

Tami Oldham and Richard Sharp were not the only people to sail into Hurricane Raymond. Chris Birchard was also on the water. He worked as a commercial fisher from Alaska. He was sailing to fish around Hawaii on his boat, the *Crescent*, which he designed and built. Birchard's wife was with him, as was someone to help sail. The *Crescent* ended up in the hurricane, tossed about just as the *Hazana* had been. The *Crescent* was damaged and spent almost a day being pushed away from shore, but Birchard finally managed to get to land. All three people survived, and Birchard spent a few weeks repairing his boat.

▲ The ability to use a sextant and nautical charts to navigate is an important skill for anyone venturing into the ocean.

was still above deck, screamed. Then the *Hazana* flipped end over end. Oldham opened her eyes. She tumbled in the yacht as it bounced on the rough seas. She smacked her head against the wall, which knocked her out.

KEEP GOING

Oldham was unconscious for 27 hours. When she awoke, she headed to the deck. The storm was gone, and so was Sharp. The safety line that had connected him to the *Hazana* was hanging over the side of the boat. The *Hazana* was a mess. Belowdecks, water filled much of the cabin. Items were strewn all over the place, some of them broken. Above deck, parts were in ruins. The masts were broken, leaving the sails hanging in the water. Important mechanical and electrical equipment was also broken, including the motor and the navigation system.

Oldham was determined to reach land. She repaired what she could so the boat could sail. She removed seawater from the vessel. Knowing San Diego was too far, Oldham aimed for Hawaii, though at 1,500 miles (2,400 km), that was a long distance to travel.[9] But she was set on surviving. She had supplies, eating peanut butter and canned food. She needed her watch to determine longitude but could not find it. However, using a sextant, some almanacs and tables, and the sun, she could navigate by latitude only.

A week into her ordeal, after bailing most of the water out of the bottom of the boat, Oldham found her watch. Now she could also calculate longitude. She said of the discovery, "Once I got my watch, it changed everything."[10]

The possibility of making an error worried her. "I ran the risk of being off the latitude of Hawaii, so that was always really heavy on my mind," Oldham explained. "If I did not get to Hawaii, I would die."[11]

At one point, while belowdecks, she heard a plane. Oldham quickly ran to the deck and shot off four flares.

She also grabbed an oar and waved it in the air. She had previously tied a red T-shirt to the end of the oar, figuring the color would attract attention. But the plane did not change its course.

Oldham struggled with doubts but did not give up. At one point, she found she had drifted off course. Determined to reach land, she decided not to sleep that night. She stayed at the wheel to keep the *Hazana* on track. She saw the lights of Hilo Harbor at 2:30 a.m. but did not get too close. In the dark, she feared hitting the large reef she knew was there.

Finally, dawn came. Just as Oldham was going to head for the harbor's entrance, she saw a ship leaving. She shot flares and waved the oar with the red T-shirt. The ship headed to the *Hazana*.

RED SKY IN MOURNING

Several years after surviving her time alone at sea, Tami Oldham self-published a book about her experience, *Red Sky in Mourning*. A publishing company later published the book with the title *Adrift: A True Story of Love, Loss, and Survival at Sea*. In 2018, Oldham's book was made into a movie called *Adrift* starring Shailene Woodley as Oldham.

One scene in the film brought Oldham to tears. As she explained, "The one scene that kind of really threw me is when Shailene is leaning over the side, putting the duct tape on the hull. Just seeing her alone, with no land in sight, with that wrecked boat—oh my gosh, it just brought me right back.... I just wept."[12]

When the ship arrived, a voice from its loudspeaker asked Oldham if she was okay. After nodding her head yes, Oldham started crying. The loudspeaker voice responded, "It's okay. We'll help you. You'll be fine."[13] The voice then asked if anyone had died. Oldham nodded yes again, acknowledging the loss of Sharp. People aboard the vessel dropped down coffee on a rope and threw Oldham an apple. The ship then towed the *Hazana* into the harbor, where members of the Coast Guard Auxiliary met Oldham.

> "One thing I realised from being in solitary confinement on a piece of floating fibreglass was how much we need human contact. Without that, when you're alone, your mind just plays all kinds of tricks on you."[15]
>
> —Tami Oldham, discussing surviving being adrift at sea in a 2019 interview

Oldham had reached Hawaii after sailing alone for 41 days. Her physical wounds were healing and her heart was broken, but she was alive. The painful, frightening experience on the *Hazana* did not keep Oldham from returning to the sea. She began wearing a special charm around her neck. When asked why she sports the sextant-shaped pendant, she explained, "It reminds me of how I got home."[14]

CHAPTER 6

A POLAR EXPEDITION GONE WRONG

In 1879, a group of adventurers set out from San Francisco, California, in the USS *Jeannette*, a US Navy ship. It was July 8, and the crew of 33 men was heading to the North Pole.[1] At the time, little was known about this part of the world. The men were going to explore this uncharted land and learn what they could. But there was another goal: to sell newspapers.

The wealthy New Yorker James Gordon Bennett Jr. financed the trip. He owned the *New York Herald*, a newspaper. He was extremely interested in the Arctic and wanted the United States to be the first country to reach the North Pole. The United States was a young

◀ James Gordon Bennett Jr. spent five years planning and preparing for the USS *Jeannette*'s expedition.

nation, and achieving his goal would help strengthen the country's reputation. Additionally, the United States had bought Alaska fairly recently, in 1867, and many people were curious about what was in that area at the top of the globe.

The *Jeannette* was initially the *Pandora*, a gunboat in the British navy. Bennett purchased the *Pandora* for the expedition to the North Pole and paid for work to prepare the ship for the voyage to the Arctic. The *Jeannette* was updated inside and out, and from top to bottom. Modifications included strengthening and reinforcing the ship, such as by adding metal beams, to deal with ice. Additional work included installing new furnaces.

Bennett used his power and influence to get backing from the US Navy for the trip. Navy personnel made up most of the crew. One of the five naval officers onboard was George Washington De Long, who commanded the *Jeannette*. He had journeyed to Arctic waters before and was excited about the assignment.

> **All these repairs were so carefully made as to give every reasonable assurance that the vessel would be able to overcome any of the ordinary perils incident to navigation in the Polar Seas.[2]**
>
> —A description of work done on the USS Jeannette *as noted in an 1880 US Navy report*

The other officers included an executive officer, a navigation officer, an engineer, and a surgeon. Special duty personnel included a boatswain to manage hull maintenance, firemen, a machinist to address any mechanical issues, and a carpenter. William Dunbar was the ice pilot, someone with experience sailing in the icy waters of the Arctic Ocean.

Two nonmilitary scientists were on board as well: Raymond Newcomb was a naturalist, and Jerome Collins was a meteorologist. As the chief meteorologist for Bennett's newspaper, Collins had the additional task of reporting on the expedition. A cook and a steward would handle feeding everyone, and two Inuit men who were hunters and dog drivers would help the crew

GEORGE WASHINGTON DE LONG

George Washington De Long was an experienced sailor and military officer when he became commander of the USS *Jeannette*. De Long was born in New York City on August 22, 1844. He attended the US Naval Academy, graduating in 1865. In 1873, he was part of an expedition to Greenland. That trip ignited his passion for the Arctic. Hampton Sides, historian and author of *In the Kingdom of Ice: The Grand and Terrible Polar Voyage of the USS* Jeannette, described De Long in a *National Geographic* interview: "He was captivated by the puzzle of what's up there at the top of the world. And he devoted the rest of his life to the goal of being the first man to reach the North Pole."[3]

▲ Ice is one of the greatest dangers of Arctic sea travel. It can puncture or trap ships.

travel over land once the ship reached its destination. The remaining crew were sailors.

Bennett was not entirely in charge of the expedition. Although he privately owned the ship, the project was a partnership with the navy, and the ship would sail under the navy's orders and rules. Bennett's wish was for his ship simply to reach the North Pole. The US Navy also wanted crew members to study the North Pole while they were there. Later, the head of the US Navy added an item to the to-do list: Look for the crew of the *Vega*, which no one had heard from for almost a year.

TRAPPED

From San Francisco, the *Jeannette* headed north starting on July 8, 1879. De Long's route was through the icy waters of the Bering Strait, the strip of water between Russia and Alaska. He planned to take the *Jeannette* to Wrangel Island, located near the coast of Siberia.

In the 1870s, people mistakenly believed Wrangel Island was much larger than it is and reached considerably farther north than it does. Because De Long thought Wrangel Island was a giant landmass, he planned to use it as a guide, following its coastline northward. De Long wanted to travel as far north as he could in the *Jeannette* before finishing the trip to the North Pole via sled.

On August 27, the *Jeannette* continued north from Siberia's Saint Lawrence Bay. The next week, on September 4, the crew spotted Herald Island.

THE BERING STRAIT

Most oceans have straits. A strait is a strip of water that connects two larger bodies of water. The Pacific Ocean has more than a dozen straits. The Bering Strait is the northernmost part of the Pacific and connects the Bering Sea and the Chukchi Sea, which is part of the Arctic Ocean. The Bering Strait separates the continents of Asia and North America. At its narrowest point, the Bering Strait is 53 miles (85 km) wide, and scientists believe people crossed from Asia to North America at this point thousands of years ago.[4]

NORDENSKIÖLD AND THE *VEGA*

Adolf Erik, Baron Nordenskiöld, set out to travel to the Bering Strait aboard the *Vega* about a year before the *Jeannette* set sail. Nordenskiöld was a Finnish-born Swedish geographer, geologist, and mineralogist. He was also an explorer. Nordenskiöld made multiple ocean voyages in the 1860s, 1870s, and 1880s, including to Greenland to study its ice. In July 1878, he left Norway on the *Vega* and headed to Siberia. Then the *Vega* got stuck in the Bering Strait for almost a year, frozen. But Nordenskiöld and the *Vega* had a good outcome. The *Vega* made it through the ice, reached Alaska, and then traveled to China, Sri Lanka, and the Suez Canal before landing in Sweden.

Shortly thereafter, the *Jeannette* got stuck in the ice and could not break free. This was not unusual when exploring near the pole, and the crew had supplies for three years. Trapped in the ice, the ship drifted for 21 months but in the correct direction, northwest. Even while stuck, team members focused on science.

In May 1881, the crew discovered two islands, which they named Henrietta and Jeannette. This made the crew hopeful. Historian Hampton Sides noted in his book *In the Kingdom of Ice: The Grand and Terrible Polar Voyage of the USS* Jeannette, "Everyone felt that the expedition, despite its grueling hardships, had made significant accomplishments—including exploring many hundreds of miles of the planet never before seen by man."[5] They learned about geography, meteorology, and ocean currents and, in the process, proved some information incorrect. For example, they learned that Wrangel Island was only an island, not a large chunk of land connected to Greenland. De Long, who kept a journal, said his crew had "exploded so many theories of other people."⁶

Finally, around midnight on June 11, the ice cracked, sending the *Jeannette* into open water. But trouble was ahead. On the afternoon of June 12, the open water began to close up again. The ice exerted a tremendous amount of pressure and started crushing the *Jeannette*. Realizing the danger of the situation, De Long had the crew move its dogs, supplies, and equipment off the ship and onto the ice. They also hauled three small boats and oars off the ship.

▼ Items the crew saved from the *Jeannette* included food, furs, guns, ammunition, medicine, navigation equipment, sleds, and harnesses.

The *Jeannette* sank the next morning, June 13. Everyone had made it off the ship. Now, the crew members had a new focus. They were no longer on a journey of discovery. They needed to survive.

ON THE MOVE

The men had made it off the ship, but they could not stay put. Rescue was highly unlikely. On June 18, they headed to Siberia, the closest land. It was nearly 1,000 miles (1,600 km) away, and they had only a few months of decent weather to make the journey.[7] The group trekked across the ice. De Long and Dunbar made their best guesses at a route.

Sometimes, the group would travel across spots of open water in their small boats. Other times, they would use a large piece of ice to tow their supplies, pulling it with ropes tied to hooks sunk into the ice. Initially, the men were mostly happy, glad to be free from the ice they had been trapped in. That feeling did not last long.

The team celebrated the Fourth of July on an ice cap in the East Siberian Sea, American flags waving from their tent tops. But the men were exhausted, in pain, and experiencing different physical problems. They were hungry. Still, the team pressed on. On September 10, the group landed

▲ The East Siberian Sea is part of the Arctic Ocean. It lies between the New Siberian Islands to the west and Wrangel Island to the east.

at Semyonovsky Island, located 100 miles (160 km) from Siberia's coast.[8] On the morning of September 12, the crew members got back into their three boats and headed toward the mainland. The men were excited by the prospect of finally reaching Siberia, but their joy was short-lived.

By the early evening, a storm blew in. The waves were high, and the waters were rough. One of the boats capsized. Its members were lost to the sea. The two remaining boats landed in different spots. De Long had captained one boat and was on the east side of the Lena River. George W. Melville, the crew's engineer, led the other group and was on the west side. The two men had very different experiences.

De Long and his group started walking inland, hoping to find a village where they might get help. The marshland was wet and cold. After eating the last dog on October 9, with no

▲ An illustration from 1882 depicted the survivors, including George W. Melville, *center left*.

more food and shelter, members of De Long's team began to succumb to starvation and exposure. De Long decided to send out a party of two men ahead of him, hoping they would find help. The pair finally encountered people, but by then it was too late for the rest of their group. De Long and the others they left behind died. Throughout the entire journey, De Long made entries in his log. De Long's last entry was on October 30, 1881: "One hundred and fortieth day. Boyd and Görtz died during night. Mr. Collins dying."[9]

Melville and his men, however, found a settlement, and the villagers gave them aid. Next, Melville headed for Bulun, Russia, where he found the two survivors from De Long's group. From Bulun, Melville went looking for De Long with the help of locals. The search party found the spot where De Long's boat had landed. There, Melville retrieved important documents from the *Jeannette*, including the ship's log, but he did not find De Long. Melville returned to Bulun, arriving on November 27. But he was not ready to give up on locating De Long.

Determined to find his captain, Melville headed north again. This time, he succeeded. Melville and the two men with him discovered De Long and the other men on March 23, 1882. He left his dead crewmates, building a cairn over their bodies. Melville then tried to find the third boat from their expedition, the one that did not survive the storm. He found no trace of it and gave up the search.

He headed to Irkutsk in Siberia, reaching the city on July 5, 1882. It was three days before the third anniversary of the start of the *Jeannette*'s expedition. The ship, her captain, and ten of her crew had been lost, but many of the men, the ship's log, and the scientific information the team gathered had survived.[10]

In 1883, the US government had the remains of De Long and the other men who died with him in Siberia moved to the United States. After a long journey through Europe, the remains were buried in Woodlawn Cemetery in New York. A monument at the US Naval Academy honors these men and the others who perished on the *Jeannette*'s journey to the North Pole.

CHAPTER 7

A RECORD-BREAKING JOURNEY

As a shark fisher, José Salvador Alvarenga had spent countless hours on the water. And when the 35-year-old from El Salvador headed out to fish off the coast of Costa Azul, Mexico, on November 17, 2012, the day was nothing out of the ordinary. Alvarenga's 25-foot (7.6 m) boat was packed and ready.[1]

He stocked the boat with drinking water, bait, hooks, line, a harpoon, knives, buckets to bail water, and ice to keep fish cool. He also had GPS equipment, a mobile phone, and tools and fuel for the motor. His plans were set. He and his fishing partner, 22-year-old Ezequiel Córdoba, planned to be gone for two days. But that is not what happened.

Alvarenga would eventually drift to the Marshall Islands in the Pacific Ocean between Hawaii and the Philippines.

LOST AT SEA

Alvarenga usually fished between 50 and 100 miles (80–160 km) offshore.[2] On November 17, he headed out to sea knowing a storm was coming, but he was not concerned. He was an experienced ocean voyager. He had dealt with bad weather numerous times. He and Córdoba had a successful fishing trip, catching mahi-mahi, sharks, and tuna. The large icebox on the boat for storing fish was full.

The men were about 50 miles (80 km) from the shore when the storm hit.[3] Alvarenga's boat had no awning to shed rainwater, and the boat began filling with water, but he did not worry. He focused on steering his craft, navigating it through the rough water toward shore. Córdoba worked frantically to get water out of the boat. He was panicky and stopped bailing at times to cling to the side of the boat, fearful. He cried, certain they were going to die.

Finally, the storm paused and the clouds broke long enough for Alvarenga to see land. He determined they were two hours from shore. That is when a new issue emerged. The boat's motor died. Alvarenga promptly called his boss, Willy, who asked for Alvarenga's coordinates, but he could not provide them. The GPS equipment did not work. Willy told him to drop anchor, but the boat did not have one.

Willy radioed he was coming to get them. Alvarenga told him to come right away.

Unable to move the boat, the two fishers were stuck and at the mercy of the sea. The storm continued and brought rough waters. Alvarenga thought the storm would last five days. The packed icebox made the boat unsteady by weighing down the front end, so he decided they should throw the haul of fish they had caught overboard to help stabilize the craft. They also discarded the motor fuel and the ice. Alvarenga had dozens of buoys on board. He strung them on a line that he attached to the back of the boat to slow down the boat, acting as an anchor of sorts.

> **It wasn't the storm that was the problem. My engine gave out.**[4]
>
> —José Salvador Alvarenga, on how he ended up adrift in the Pacific Ocean

Alvarenga was doing what he could to keep them afloat until his boss arrived. But then the radio stopped working. He had no way of communicating with Willy. Alvarenga and Córdoba were alone.

The storm battered them. The two men worked hard to scoop water out of the boat. Their bodies were exhausted, and the night brought cold temperatures. Alvarenga's boat

ANDREA GAIL

For centuries, inhabitants of Gloucester, Massachusetts, have taken to the sea to fish, and many people have lost their lives in the pursuit. One example happened in October 1991. The *Andrea Gail*, a 70-foot (21 m) boat, had traveled 900 miles (1,450 km) to Newfoundland, Canada, to catch swordfish. On its return, it was caught in a fierce storm ravaging the East Coast. The *Andrea Gail* encountered winds of 120 miles per hour (190 kmh).[5] The boat's owner began to worry when he did not hear from the boat for three days. The Coast Guard searched for the vessel and survivors for ten days but found nothing. The 2000 movie *The Perfect Storm* is based on this story.

provided no shelter. The two were exposed to the elements. What they did have was the large icebox for storing fish. They flipped it over and huddled inside for shelter and to get warm. The storm continued to fill the boat with water. Alvarenga and Córdoba took turns leaving the icebox to bail water, working a few minutes at a time. All the while, the storm pushed them farther away from shore.

The storm lasted seven days. The turbulent water thrashed them about and tossed Córdoba overboard. But Alvarenga was able to get him back in the boat, pulling the young man by his hair.

Eventually, the waters calmed. The two men had survived the storm, but they had a new dilemma. They were far from where they had started and lost at sea. They had no

▲ Alvarenga's boat offered no protection from the elements while the two men were adrift at sea.

provisions, no radio, and no flares. They were in a tiny boat on the massive Pacific Ocean.

TRYING TO SURVIVE

With Alvarenga and Córdoba unable to meet their basic needs, surviving was going to be extremely difficult. Alvarenga explained, "We didn't think about hunger at first. It was the thirst. We had to drink our own urine after the storm."[6]

Even without fishing equipment, which had been lost during the storm, Alvarenga was able to catch a few fish. He had been fishing since childhood and knew how to catch fish with his bare hands. He would throw the fish in the boat, and Córdoba would clean and fillet them. Despite the fish and an occasional turtle, the two men did not have enough food. They were also desperate for water. Sometimes they encountered garbage that could provide sustenance. At one point, they happened upon carrots, part of a cabbage, and a container of milk that had turned sour. They welcomed all of it.

Alvarenga and Córdoba suffered from lack of shelter as well. The sun beat down on them. They decided to climb into the icebox for cover.

Soon, the boat had visitors. Seabirds landed, taking advantage of the craft to rest from their long flights. Alvarenga took advantage as well. He grabbed a bird when

▼ Triggerfish were among the fish that Alvarenga and Córdoba relied on for food.

he could, killed it, and then he and Córdoba consumed it. They ate almost every part of the bird, even feathers. Desperate for nourishment and water, they also drank the blood.

Alvarenga and Córdoba ate what they could find in the ocean, such as seaweed. But their sources of food remained inconsistent. Alvarenga did his best not to panic, but Córdoba struggled to remain calm. Adrift for two months, Córdoba began to give up. Eating the raw meat from birds had made him ill, so he decided to stop eating. Alvarenga described Córdoba's state of being:

> *He would cry a lot, talking about his mama, eating tortillas, and drinking something cold. I helped him as much as I could. I would hug him. I told him, "We'll be rescued soon. We'll hit an island soon." But he would sometimes get violent, screaming that we were going to die.*[7]

The men promised each other to visit the other person's parents if only one of them survived. Córdoba would go to El Salvador to see Alvarenga's mother and father, and Alvarenga would visit Córdoba's mother in Mexico.

Several weeks into their ordeal, Córdoba's condition worsened. The idea of losing Córdoba frightened Alvarenga, who screamed at his boatmate, "Don't leave me alone! You have to fight for life! What am I going to do here alone?"[8] The men were huddled in the box they had been using for shelter. Minutes later, Córdoba died. According to Alvarenga, "We said our goodbyes. He wasn't in pain. He was calm. He didn't suffer."[9]

Initially, Alvarenga kept Córdoba's body on the boat. He even began to speak to the corpse. "Buenos días," Alvarenga greeted his companion. Then he asked, "What is death like?" Alvarenga answered on Córdoba's behalf, "Good. It is peaceful."[10] Alvarenga wanted peace too, and he thought about killing himself. After several days of talking to Córdoba's lifeless body, Alvarenga thought he was losing his mind and decided it was enough. He gave Córdoba's body to the ocean.

NEVER GIVE UP

Alvarenga was not ready to give up. He focused harder on finding food. He prayed. He sang hymns. When multiple cargo ships passed by without helping him, he did not lose hope. "I would signal them and nothing would happen," Alvarenga explained. "But I thought God will determine which boat will save me."[11]

Alvarenga also used his imagination to keep his spirits up. He let his mind take him somewhere nice. Each morning, he went for a walk. "I would stroll back and forth on the boat and imagine that I was wandering the world," he said. "By doing this I could make myself believe that I was actually doing something. Not just sitting there, thinking about dying."[12]

On January 29, 2014, Alvarenga drifted near land. When he was close enough to swim ashore, he did. He had reached one of the islands in the Marshall Islands. He had drifted more than 6,000 miles (9,700 km).[13] Alvarenga described the experience: "I felt the waves, I felt the sand, and I felt the shore. I was so happy that I fainted on the sand."[14]

Alvarenga found people not far from where he swam ashore. They helped him immediately. Alvarenga had

DISBELIEVERS

Following his rescue, José Salvador Alvarenga became a news sensation. Reporters hounded him. Some sneaked into the hospital where he was recovering. Some were on his flight home and tried to photograph him. Many reporters were skeptical of Alvarenga and questioned whether he was telling the truth. He ignored them. Several oceanographers, who are scientists specializing in ocean studies, were not skeptical. They declared that what happened to Alvarenga was possible. Data from buoys and weather models confirmed a person lost at sea where he was could indeed drift thousands of miles across the Pacific Ocean and land at the Marshall Islands where Alvarenga went ashore.

▲ A sea patrol boat brought Alvarenga from the Ebon Atoll, where he landed, to Majuro, the capital city of the Marshall Islands, where he could receive treatment.

survived being lost at sea for 438 days. He has the record for surviving the longest while adrift on the ocean.

Following his rescue, Alvarenga spent time in a hospital. Then he flew to El Salvador. After returning to his homeland, he followed through on his promise to Córdoba and visited his mother in Mexico. Two years following his rescue, Alvarenga had not yet returned to the water. He stopped fishing, scared of the sea. In an interview with CNN in January 2016, he shared, "There are still nights when I can't sleep. The ocean keeps haunting me."[15]

> "I'm happy to be alive. I'm happy to be with my family. I'm proud to be what I am. I am simply glad I'm here."[16]
>
> —José Salvador Alvarenga in an interview following his return home

CHAPTER 8

SURVIVING OCEANS

Surviving being lost at sea involves a bit of luck. But people can take a variety of precautions to improve their chances of returning home alive should they end up stuck in the ocean. When heading out to sea, people should hope for the best and prepare for the worst.

Preparing means bringing gear with survival in mind. Grenada Bluewater Sailing, which has a sailing school and a Royal Yachting Association training center, recommends taking a grab bag, describing it as a "vital piece of safety equipment that must be prepared and ready to go in case of an emergency."[1] The bag should be waterproof and able to float.

◀ The ocean is a beautiful place, and many people enjoy exploring its wonders. Visitors should take precautions to help them stay safe.

Essential items to keep in a grab bag include a passport, money, the ship's log or papers, and a very high frequency radio to call for help. They also include drinking water or a filter that can remove salt and contaminants, food, flares, a mirror, a first aid kit, and a whistle to attract attention. Navigation tools, a flashlight with batteries, oars, and a sea anchor—which is like a parachute that drags in the water to slow the boat's movement—are helpful during emergency travel. Additional items might include a fishing kit or speargun, a water maker, a life raft repair kit, and sunscreen.

A life raft is important. Steven Callahan would not have survived 76 days at sea without his raft. Investing in a good raft, which can come with paddles, buckets, a covered deck, pouches for collecting water, flares, and more, can be well worth the expense. But even a simple raft can be enough to help a person survive. The most important thing is that it is on the boat and it floats.

People should also be prepared to survive if they end up in cold water, such as if their boat capsizes. People should dress in layers of clothing, including a waterproof layer if possible. The layers will help the person stay warm longer. Wearing a life jacket is also important. It will keep the head above water, including when the wearer becomes

▲ Callahan demonstrated a life raft he designed with a sail, which he called the Clam.

tired and desperate. Of the 477 US drowning deaths of recreational boaters in 2022, 85 percent of those who died were not wearing a life jacket.[2]

A harness is another safety feature to keep in mind. A harness with a tether can keep a person secured to a boat. If they get swept overboard, the tether can keep them with the boat so they don't get lost at sea.

Gear includes electronic equipment. GPS units are valuable tools for people navigating land and sea. But they can be lost overboard or short out from water splashing on board. They can also lose power. Whatever the reason, losing a GPS device can make navigating hard to do. If this happens, it is important to have another tool available as a backup.

Both Callahan and Oldham used a sextant to navigate. Oldham expressed how important it was to her survival:

> ### THE CLAM
> Following his experience of being adrift at sea for 76 days in a raft, Steven Callahan designed a life raft for others who might face the same ordeal. He calls his design the Clam, saying that it works better than a typical life raft. Like his *Rubber Ducky III*, the Clam has a canopy to provide shelter from the sun and rain. But the Clam differs from the *Rubber Ducky III* in important ways. Callahan's design features a fiberglass bottom, making it harder and more durable than nylon, and a sail, which allows a person to move the craft intentionally rather than be stuck adrift.

"Even if it's a plastic sextant—I always say that if you're going to do ocean crossings, make sure there's a sextant on board with the relevant tables. Because in the end, if all the batteries go, you have that to fall back on. And that is what saved my life."[3]

▼ Life jackets are among the most important things to bring along on a boating excursion.

▲ High winds from storms can cause waves that could capsize boats.

WEATHER

Because rough and stormy seas can overturn or damage a boat, knowing the weather is important. It could mean the difference between life and death. Oldham learned this firsthand when she and Richard Sharp sailed directly into Hurricane Raymond in 1983. She lost her fiancé, was injured, and was stuck in the Pacific Ocean on a broken yacht.

In her 2019 Royal National Lifeboat Institution interview, Oldham provided advice for readers. Her first point was about the weather: "Don't mess with Mother Nature. Study the weather, because Mother Nature is much bigger than we are. That's one of the most important things I can say."[4]

SHELTER AND CLOTHING

Having shelter can provide protection against wind, rain, sunlight, and ocean water. Shelter can come in the form of a cabin belowdecks, such as the one Oldham escaped to during Hurricane Raymond. It can also be a simple canopy like the one on Callahan's raft. People can also create makeshift shelters, such as the large icebox for storing fish on Alvarenga's boat.

Clothing also serves as protection against the elements. Worn in layers, clothing can provide warmth in the cold.

During the day, pieces of clothing can provide shade when placed overhead.

WATER AND FOOD

Having provisions may help those stuck at sea stay alive long enough to be rescued. Water is most important. A person can survive longer without food than without water. Because drinking seawater can speed up dehydration, not drinking water from the ocean is critical.

For those who have drinking water available, rationing is important. People can never know how long they will be adrift before being rescued or coming across land. Capturing rainwater will provide drinkable water. A variety of objects can hold water. Options include buckets, the hood of a raincoat, and even watertight boots. Clothing is an option too. Soaked fabric can be wrung to release the water. Bringing a tarp to catch rainwater and bottles to store the water is good planning.

Because objects will have been exposed to seawater, they should be rinsed off by rain before serving as rain collection vessels. But even the saltwater can be useful. Rich Johnson wrote in *The Ultimate Survival Manual*, "The first water you collect will have a high salt content, so store it separately, and use it to clean wounds or to wash food before eating."[5]

Oldham had canned food available during her 41 days at sea. Callahan and Alvarenga were not so fortunate, but they did catch an occasional fish and other ocean animals. This helped keep them alive long enough for rescue.

Fish are drawn by the shadow cast by a boat. If a fishing rod is not available, everyday objects can take its place. For example, a shoelace can serve as a fishing line. To attract fish, people can add shiny jewelry to act as a lure. A feather could also serve as a lure. If fishing with a hook from an inflatable raft, it is critical to avoid puncturing the raft with the hook. After catching a fish, keeping uneaten pieces can prove useful. They can be bait for the next round of fishing.

STRANDED IN THE WATER

Not everyone who experiences trouble at sea ends up in a vessel. Some people find themselves in the water. Ideally, a person in this situation will be wearing a wet suit to help keep them warm and a life vest to keep them afloat. Pulling the knees to the chest and holding that position can help retain body heat.

People in open water face greater challenges than those in a boat or raft. In the open water, they cannot collect drinkable water or catch fish or birds to eat. These people are also more vulnerable to sea creatures, including sharks. With all the challenges being in the open water presents, survival beyond a few days is unlikely.

RICHARD VAN PHAM

In May 2002, 62-year-old Richard Van Pham headed out to cruise the ocean waters near Long Beach, California, in his 26-foot (8 m) boat, the *Sea Breeze*. What he expected to be a journey of a few hours ended up going for 2,500 miles (4,020 km) and four months. He hit a storm that broke the mast of his boat. Next, the motor and radio died. He survived by collecting rainwater and catching fish and a turtle. A US government plane spotted the *Sea Breeze* 300 miles (480 km) from Costa Rica and sent a nearby US Navy vessel to investigate.[6] When found, Van Pham only wanted his boat repaired, but the sailors who found him said it was beyond repair. They sank his boat and took him ashore.

FIGHTING SUN EXPOSURE

Limiting sun exposure is good for multiple reasons. Getting hot increases sweating, which leads to rapid water loss and the possibility of dehydration. To keep the body cooler, people should stay in the shade. If needed, they can create shade by placing fabric such as a tarp, a sail, or even clothing overhead. They should be careful not to cut off airflow, which could make things hotter rather than cooler.

Another way to cool off is with water. Splashing it on the body can help. Going into the ocean is an option, but tying oneself to the raft or boat is a must. Keeping an eye out for sharks is also crucial. Putting one's hands and feet in cold water is a good method for cooling down quickly if getting into the water isn't an option.

⚠️ **A wet suit can help keep a person warm and dry and also protect against sunburn.**

Wet clothing can be particularly helpful in dealing with heat. If wearing wet clothing to stay cool, people should let it dry off before sundown, when temperatures drop. Wet clothing could make a person become too cold at night.

Sunlight damages the skin. Applying sunscreen is important. If none is available, clothing can protect against sunburn.

ATTRACTING ATTENTION

Ships or aircraft may pass nearby or overhead when a person is stranded in the ocean. Having a way to attract these passersby could aid in being rescued. A variety of tools can meet that need.

Flares are designed with rescue in mind. Other types of signaling devices are audible, electronic, or visual. Audible options include an air horn and a whistle. An item that is visual and electronic is a beacon that runs on batteries and flashes the SOS distress signal. People can also wave a small distress flag.

Everyday objects can also prove helpful. Using a smartphone screen or mirror to reflect sunlight can catch the eye of someone as far as 10 miles (16 km) away.[7]

NOT GIVING UP

With all the items one can take on an ocean trip, one of the most vital is already with them. It is mental strength. It keeps a person going in the face of seemingly impossible adversity.

Callahan, Okene, Oldham, Alvarenga, and the survivors of the USS *Jeannette* had different ocean ordeals, but all of them faced the same challenge. They had to not give up. And that was not easy. But amid their struggles, they managed to keep going. Without that determination, they would not have survived.

The ocean environment is vast and varied. It serves various purposes for different people. It offers great possibilities for learning, employment, adventure, and fun. But when things do not go as planned or expected, the ocean can also become a terrifying place.

As many real-life stories reveal, mishaps while traveling the seas can get the best of even the most seasoned sailors. But numerous stories also show that survival is possible in worst-case scenarios. These examples can help others be prepared and increase their own odds of survival if they too end up stranded at sea.

> "To this day I feel enlightened by what I went through because it changed me for the better. But would I want to be adrift in the ocean again? No way."[8]
>
> —Steven Callahan in a 2012 article in the *Guardian*

ESSENTIAL FACTS

SURVIVAL STORIES

- While Steven Callahan was sailing from Europe to the United States, something struck his sailboat, the *Napoleon Solo*, causing it to fill with water. Callahan escaped to his life raft, where he spent 76 days adrift on the Atlantic Ocean.

- When the tugboat Harrison Okene worked on was overturned by a huge wave, he sank to the bottom of the ocean with the boat, alive thanks to an air bubble. He survived more than two days before divers rescued him. Okene was the only survivor from his crew.

- Tami Oldham was sailing for California from Tahiti with her fiancé, Richard Sharp, when the couple sailed into a hurricane. Oldham survived and managed to reach Hawaii after 41 days adrift on the Pacific Ocean in a battered yacht, but Sharp was lost at sea.

- In 1879, the USS *Jeannette* and its crew set out from California on an expedition to explore the North Pole. After being stuck in ice for many months, the ship sank two years into the journey. The crew was divided and left to find help along the frigid Bering Strait. Several men died, including the captain, but some of the crew survived.

- José Salvador Alvarenga was an experienced fisher who set out to fish off the coast of Mexico for two days with Ezequiel Córdoba. When Alvarenga's motor stopped working and a storm came, he ended up adrift for more than a year, during which time Córdoba died. Alvarenga drifted thousands of miles across the Pacific before he finally reached land.

OCEAN SURVIVAL

- Earth has a single ocean that covers more than 70 percent of the planet.

- The ocean is the world's largest ecosystem and is home to millions of species of plants and animals.

- Bringing equipment such as communication, flotation, and signaling devices can improve one's chances of being rescued.

- Because rough seas can overturn or damage a boat, knowing the weather forecast is important.

- Having drinkable water is essential to survival. People should not drink ocean water; its salt content means it will speed up dehydration.

- People should never give up hope of surviving.

QUOTE

"I wish I could describe the feeling of being at sea, the anguish, frustration, and fear, the beauty that accompanies threatening spectacles, the spiritual communion with creatures in whose domain I sail."

—*Steven Callahan, written in Bermuda in 1981*

GLOSSARY

bail
To remove water from a boat by scooping it out and pouring it over the side of the boat.

boatswain
A ship's officer responsible for keeping smaller boats, sails, rigging, and other equipment in order.

cairn
A mound of stones built as a landmark.

capsize
To flip over, as a boat.

coma
Unconsciousness due to disease, injury, or poison.

ecosystem
A community of interacting organisms and their environment.

hull
The frame or body of a ship, excluding masts, sails, and rigging.

insurance
Coverage people pay for that will give them money to help replace what has been damaged or destroyed in case they experience a loss, such as property loss from a storm or crash.

latitude
A distance north or south of the equator, measured in degrees.

longitude
A position along any of a series of imaginary lines running from north to south on the globe, indicated by degrees.

mast
The tall pole in the center of a boat that holds rigging and a sail.

provision
A supply of something, such as food.

rigging
The ropes and chains on a boat, especially related to the sails.

sloop
A sailboat that has one mast.

SOS
A call for help that has its origin in Morse code, communicated in a sequence of three long signals, three short signals, and three long signals.

yacht
A large boat with a motor and sails used for cruising or racing.

ADDITIONAL RESOURCES

SELECTED BIBLIOGRAPHY

Ashcraft, Tami Oldham. *Adrift: A True Story of Love, Loss, and Survival at Sea*. William Morrow, 2018.

Callahan, Steven. *Adrift: 76 Days Lost at Sea*. Houghton Mifflin, 1999.

Ocean: The Definitive Visual Guide. DK, 2022.

FURTHER READINGS

McKinney, Donna B. *US Coast Guard*. Abdo, 2021.

Piven, Joshua, and David Borgenicht. *The Worst-Case Scenario Survival Handbook: Expert Advice for Extreme Situations*. Chronicle Books, 2019.

Tougias, Michael J., and Alison O'Leary. *Abandon Ship! The True World War II Story about the Sinking of the Laconia*. Little, Brown, 2023.

ONLINE RESOURCES

To learn more about the ocean and survival, please visit **abdobooklinks.com** or scan this QR code. These links are routinely monitored and updated to provide the most current information available.

MORE INFORMATION

For more information on this subject, contact or visit the following organizations:

BISCAYNE NATIONAL PARK FLORIDA
9700 SW 328th St.
Sir Lancelot Jones Way
Homestead, FL 33033
nps.gov/bisc/index.htm

One of several oceanic national parks in the United States, Biscayne National Park Florida is 95 percent water and can be explored while boating, canoeing, kayaking, and snorkeling.

MONTEREY BAY AQUARIUM
886 Cannery Row
Monterey, CA 93940
montereybayaquarium.org

The Monterey Bay Aquarium's more than 200 exhibits feature 80,000 ocean plants and animals, including jellyfish, penguins, sea otters, and sharks. The organization also provides learning opportunities online.

SEA SCOUTS BSA
1325 W. Walnut Hill Ln.
PO Box 152079
Irving, TX 75015
seascout.org

The Sea Scouts BSA program is intended for youth ages 14 to 20 years old and focuses on boating, water skills, and survival.

SOURCE NOTES

CHAPTER 1. STEVEN CALLAHAN, SEA LOVER

1. "Manry, Robert N." *Encyclopedia of Cleveland History*, n.d., case.edu. Accessed 2 Aug. 2023.
2. Joe Cline. "Steve Callahan: Part Two." *48° North*, 17 Dec. 2020, 48north.com. Accessed 2 Aug. 2023.
3. Emilie Tavel Livezey. "Survival at Sea." *Christian Science Monitor*, 30 Sept. 1982, csmonitor.com. Accessed 2 Aug. 2023.
4. "Real-Life Shipwreck Survivor Helped *Life of Pi* Get Lost at Sea." *NPR*, 21 Feb. 2013, npr.org. Accessed 2 Aug. 2023.
5. Steven Callahan. *Adrift: 76 Days Lost at Sea*. Houghton Mifflin, 1999. xvi.
6. Callahan, *Adrift*, xvi
7. "Manry, Robert N."
8. Callahan, *Adrift*, xviii.
9. Callahan, *Adrift*, xvii.
10. Steven Callahan. "Experience: I Was Adrift on a Raft in the Atlantic for 76 Days." *Guardian*, 23 Mar. 2012, theguardian.com. Accessed 2 Aug. 2023.
11. Livezey, "Survival at Sea."

CHAPTER 2. THE DANGERS OF THE OCEAN

1. Alyn C. Duxbury and Claudia Cenedese. "Ocean." *Encyclopedia Britannica*, 28 July 2023, britannica.com. Accessed 2 Aug. 2023.
2. "Voyager: How Long Until Ocean Temperature Goes up a Few More Degrees?" *UC San Diego Scripps Institution of Oceanography*, 18 Mar. 2014, scripps.ucsd.edu. Accessed 2 Aug. 2023.
3. "What Is a Thermocline?" *National Ocean Service*, 20 Jan. 2023, oceanservice.noaa.gov. Accessed 2 Aug. 2023.
4. *Ocean: The Definitive Visual Guide*. DK, 2022. 34.
5. "What Are the Oldest Living Animals in the World?" *National Ocean Service*, 20 Jan. 2023, oceanservice.noaa.gov. Accessed 2 Aug. 2023.
6. *Ocean*, 207.
7. "Marine Life Encyclopedia: Ocean Fishes." *Oceana*, n.d., oceana.org. Accessed 2 Aug. 2023.
8. "Great Barrier Reef Facts." *Great Barrier Reef*, n.d., greatbarrierreef.org. Accessed 2 Aug. 2023.
9. Charles W. Bryant. "How Long Can You Survive Adrift in the Ocean?" *MapQuest Travel*, 12 Apr. 2021, mapquest.com. Accessed 2 Aug. 2023.
10. "Heat Stress—Heat Related Illness." *Centers for Disease Control and Prevention*, 13 May 2023, cdc.gov. Accessed 2 Aug. 2023.
11. "Hypothermia." *Mayo Clinic*, 5 Mar. 2022, mayoclinic.org. Accessed 2 Aug. 2023.
12. "Hypothermia."
13. Gavin Naylor, "The ISAF 2022 Shark Attack Report." *Florida Museum*, n.d., floridamuseum.ufl.edu. Accessed 2 Aug. 2023.
14. "Coast Guard Releases Summary of 2022 Recreational Boating Statistics." *United States Coast Guard News*, 17 May 2023, news.uscg.mil. Accessed 2 Aug. 2023.

CHAPTER 3. STEVEN CALLAHAN, SURVIVOR

1. Carol Brissie. "Engrossing Tale of Shipwreck, Survival, and Rescue at Sea." *Christian Science Monitor*, 31 Jan. 1986, csmonitor.com. Accessed 2 Aug. 2023.
2. Brissie, "Rescue at Sea."
3. Ryan Sargent. "All the Ingenious & Desperate Tactics Steven Callahan Used to Survive 76 Days Stranded in Open Water." *Ranker*, 22 Oct. 2019, ranker.com. Accessed 2 Aug. 2023.
4. "Water: How Much Should You Drink Every Day?" *Mayo Clinic*, 12 Oct. 2022, mayoclinic.org. Accessed 2 Aug. 2023.
5. Steven Callahan. *Adrift: 76 Days Lost at Sea*. Houghton Mifflin, 1999. 41.
6. Callahan, *Adrift*, 51.
7. Callahan, *Adrift*, 61.
8. Steven Callahan. "Experience: I Was Adrift on a Raft in the Atlantic for 76 Days." *Guardian*, 23 Mar. 2012, theguardian.com. Accessed 2 Aug. 2023.
9. Callahan, "Adrift on a Raft."
10. Brissie, "Rescue at Sea."
11. Callahan, "Adrift on a Raft."
12. Tammy Oaks. "Sinking Survivor Designs Life Raft." *CNN*, 22 Apr. 2002, edition.cnn.com. Accessed 2 Aug. 2023.
13. Callahan, *Adrift*, 194.
14. Emilie Tavel Livezey. "Survival at Sea." *Christian Science Monitor*, 30 Sept. 1982, csmonitor.com. Accessed 2 Aug. 2023.
15. "Adrift: 76 Days Lost at Sea: Description." *Broomfield Public Library*, n.d., broomfield.marmot.org. Accessed 2 Aug. 2023.
16. "Adrift: 76 Days Lost at Sea: About the Author." *Amazon*, n.d., amazon.com. Accessed 2 Aug. 2023.
17. Brissie, "Rescue at Sea."
18. Mikey O'Connell. "Oscars: *Life of Pi* Tops with 4 Wins; *Argo* Named Best Picture." *Hollywood Reporter*, 24 Feb. 2013, hollywoodreporter.com. Accessed 2 Aug. 2023.
19. "Real-Life Shipwreck Survivor Helped *Life of Pi* Get Lost at Sea." *NPR*, 21 Feb. 2013, npr.org. Accessed 2 Aug. 2023.

CHAPTER 4. TRAPPED UNDERWATER

1. Raffaella Ciccarelli. "The Man Who Survived Two-and-a-Half Days Trapped on the Bottom of the Atlantic Ocean." *9News*, 25 July 2022, 9news.com.au. Accessed 2 Aug. 2023.
2. Ciccarelli, "Two-and-a-Half Days Trapped."
3. Jess Thomson. "Man Found Alive at Bottom of Sea 3 Days after Boat Sank: 'Total Shock.'" *Newsweek*, 14 Jan. 2023, newsweek.com. Accessed 2 Aug. 2023.
4. Thomson, "Found Alive at Bottom of Sea."
5. "Great Survival Stories: Harrison Okene, the Accidental Aquanaut." *Explorersweb*, 5 Mar. 2021, explorersweb.com. Accessed 2 Aug. 2023.
6. Ciccarelli, "Two-and-a-Half Days Trapped."
7. Ciccarelli, "Two-and-a-Half Days Trapped."
8. Ciccarelli, "Two-and-a-Half Days Trapped."

SOURCE NOTES CONTINUED

CHAPTER 5. ADRIFT, HEARTBROKEN, AND ALONE

1. "The Distance from Tahiti to San Diego, California." *Travelmath*, n.d., travelmath.com. Accessed 2 Aug. 2023.
2. "Adrift: Surviving a Hurricane at Sea." *Royal National Lifeboat Institution*, 4 Apr. 2019, rnli.org. Accessed 2 Aug. 2023.
3. Tami Oldham Ashcraft. *Adrift: A True Story of Love, Loss, and Survival at Sea*. William Morrow, 2018. 4.
4. Ashcraft, *Adrift*, 23.
5. Ashcraft, *Adrift*, 25.
6. Ashcraft, *Adrift*, 26.
7. Julie A. Jacob. "Coping on Land after Surviving at Sea." *Chicago Tribune*, 19 Mar. 2003, chicagotribune.com. Accessed 2 Aug. 2023.
8. Ashcraft, *Adrift*, 27.
9. Jacob, "Coping on Land."
10. "Surviving a Hurricane at Sea."
11. Shawnté Salabert. "The Real Survival Story behind *Adrift*." *Outside*, 7 June 2018, outsideonline.com. Accessed 2 Aug. 2023.
12. Salabert, "Real Survival Story."
13. Ashcraft, *Adrift*, 180.
14. Jacob, "Coping on Land."
15. "Surviving a Hurricane at Sea."

CHAPTER 6. A POLAR EXPEDITION GONE WRONG

1. Simon Worrall. "The Hair-Raising Tale of the U.S.S. *Jeannette*'s Ill-Fated 1879 Polar Voyage." *National Geographic*, 25 Sept. 2014, nationalgeographic.com. Accessed 2 Aug. 2023.
2. "A Lengthy Deployment: The *Jeannette* Expedition in Arctic Waters as Described in Annual Reports of the Secretary of the Navy, 1800–1884." *Naval History and Heritage Command*, 7 Nov. 2017, history.navy.mil. Accessed 2 Aug. 2023.
3. Worrall, "*Jeannette*'s Ill-Fated Voyage."
4. John Misachi. "Bering Strait." *WorldAtlas*, 3 Mar. 2021, worldatlas.com. Accessed 2 Aug. 2023.
5. Hampton Sides. *In the Kingdom of Ice: The Grand and Terrible Polar Voyage of the USS Jeannette*. Doubleday, 2014. 225.
6. Sides, *In the Kingdom of Ice*, 225.
7. Sides, *In the Kingdom of Ice*, 231.
8. Sides, *In the Kingdom of Ice*, 302.
9. Sides, *In the Kingdom of Ice*, 389.
10. "*Jeannette* I (Steam Bark)." *Naval History and Heritage Command*, 22 July 2015, history.navy.mil. Accessed 2 Aug. 2023.

CHAPTER 7. A RECORD-BREAKING JOURNEY

1. Jonathan Franklin. "Lost at Sea: The Man Who Vanished for 14 Months." *Guardian*, 7 Nov. 2014, theguardian.com. Accessed 2 Aug. 2023.
2. Franklin, "Lost at Sea."
3. Franklin, "Lost at Sea."
4. Kyung Lah. "Real-Life Castaway Survived 438 Days Lost at Sea." *CNN*, 20 Jan. 2016, cnn.com. Accessed 2 Aug. 2023.
5. Meagan McGinnes. "25 Years Ago, the Crew of the *Andrea Gail* Was Lost in the 'Perfect Storm.'" *Boston Globe*, 29 Oct. 2016, boston.com. Accessed 2 Aug. 2023.
6. Lah, "Real-Life Castaway."
7. Lah, "Real-Life Castaway."
8. Franklin, "Lost at Sea."
9. Lah, "Real-Life Castaway."
10. Lah, "Real-Life Castaway."
11. Lah, "Real-Life Castaway."
12. Franklin, "Lost at Sea."
13. Avery Thompson. "9 Unbelievable True Stories about People Who Survived Being Lost at Sea." *Popular Mechanics*, 9 July 2021, popularmechanics.com. Accessed 2 Aug. 2023.
14. Lah, "Real-Life Castaway."
15. Lah, "Real-Life Castaway."
16. Lah, "Real-Life Castaway."

CHAPTER 8. SURVIVING OCEANS

1. "What to Put in a Grab Bag." *Grenada Bluewater Sailing*, n.d., grenadabluewatersailing.com. Accessed 2 Aug. 2023.
2. *2022 Recreational Boating Statistics*. US Department of Homeland Security and US Coast Guard, 25 May 2023, uscgboating.org. Accessed 2 Aug. 2023.
3. "Adrift: Surviving a Hurricane at Sea." *Royal National Lifeboat Institution*, 4 Apr. 2019, rnli.org. Accessed 2 Aug. 2023.
4. "Surviving a Hurricane at Sea."
5. Rich Johnson. "Stranded at Sea? Follow This 7-Step Extreme Survival Guide." *CNN*, 11 Apr. 2013, cnn.com. Accessed 2 Aug. 2023.
6. Jessica Garrison and Lee Romney. "Rescued after 4 Months Adrift." *SFGATE*, 6 Oct. 2022, sfgate.com. Accessed 2 Aug. 2023.
7. Jia You. "Lost at Sea? Survive with These Tricks." *Popular Science*, 4 June 2014, popsci.com. Accessed 2 Aug. 2023.
8. Steven Callahan. "Experience: I Was Adrift on a Raft in the Atlantic for 76 Days." *Guardian*, 23 Mar. 2012, theguardian.com. Accessed 2 Aug. 2023.

INDEX

Adrift (film), 60
Adrift: 76 Days Lost at Sea, 42
Alaska, 57, 64, 67, 68
Alvarenga, José Salvador, 75–85, 93, 95, 98
Andrea Gail, 78
animals, 20–21, 95
    birds, 21, 39, 41, 80–81, 95
    dolphins, 9, 21
    fish, 11, 14, 20–22, 27–29, 32, 34–36, 38, 39, 41, 47, 75–79, 93, 95, 96
    sharks, 11, 14, 22, 27–29, 35–36, 75–76, 95, 96
    turtles, 21, 79, 96
    whales, 9, 11, 21, 54
Arctic, 17, 63–65, 67
Atlantic Ocean, 5–9, 17, 31, 33, 38, 40, 45–46

Bennett, James Gordon, Jr., 63–66
Bering Strait, 67, 68
Bullimore, Tony, 46

Callahan, Steven, 5–14, 31–42, 43, 88, 90, 93, 95, 98, 99
Coast Guard, US, 29, 34, 61, 78
Córdoba, Ezequiel, 75–82, 85
currents, 17, 19–21, 69

DCN Diving, 48–49, 51
De Long, George Washington, 64, 65, 67, 69–73
dehydration, 22–23, 94, 96
divers, 48, 50–51
diving bells, 49–51
drinking water, 14, 22, 32–35, 40, 47, 75, 79, 81, 88, 94, 95

food, 10, 12, 14, 20, 31, 34–35, 38, 39, 41, 47, 59, 72, 80–82, 88, 94–95

GPS, 75–76, 90
grab bag, 87–88

Hawaii, 57, 59, 61
heat exhaustion, 24–26
heatstroke, 24–26
hypercapnia, 49
hypothermia, 26–27, 46

ice, 19, 64–65, 67–70, 75, 77

Jascon-4, 45–46, 48, 51

life jackets, 88–90
life rafts, 11–13, 28, 31, 88, 95–96
    Clam, 90
    Rubber Ducky III, 31, 33, 35–37, 38, 40, 41–42, 90, 93

Manry, Robert, 5, 7
Marshall Islands, 83
Melville, George W., 71–73
Mexico, 75, 81, 85
Mini Transat, 9–10

Napoleon Solo, 6–7, 10–13, 31–32, 34, 40
Navy, US, 63–64, 66, 96
North Pole, 63–64, 65, 66–68, 73

Okene, Harrison, 45–51, 98
Oldham, Tami, 53–54, 56–61, 90, 93, 95, 98
oxygen, 46

Pacific Ocean, 17, 53–54, 67, 77, 79, 83, 93
Perfect Storm, The, 78

radios, 54, 56, 77, 79, 88, 96
rain, 18, 19, 40, 41, 55, 76, 90, 93–94, 96
Red Sky in Mourning, 60
Russia, 67, 72
    Siberia, 67, 68, 70–71, 73

salt, 14, 18–19, 23–24, 33, 88, 94
Sea Scouts, 6
sextants, 38, 59, 61, 90–91
Sharp, Richard, 53–58, 61, 93
Sides, Hampton, 65, 68

signals, 46, 82, 98
    flares, 12, 32–33, 39, 59–60, 79, 88, 98
Southern Ocean, 17, 46
stills, 14, 32, 34, 40
storms, 46, 55–56, 58, 71, 73, 76–79, 93, 96
    Hurricane Raymond, 55–57, 65, 93
sun, 13, 18, 21, 24, 38, 55–56, 59, 80, 88, 90, 93, 96–98

tethers, 57, 90
Tinkerbelle, 5, 7

USS *Jeannette*, 63–64, 65, 67, 68–70, 72–73, 98

van Heerden, Nico, 48–49
Van Pham, Richard, 69
Vega, 66, 68
Vendée Globe, 46

waves, 11, 28, 31, 45, 57, 71, 83
wet suits, 46, 95
wind, 10–11, 20, 54–57, 78, 93
Wrangel Island, 67–69

yachts, 46, 53, 93
    Hazana, 53, 55–58, 60–61

ABOUT THE AUTHOR

REBECCA ROWELL

Rebecca Rowell has put her degree in publishing and writing to work as an editor and as an author, working on dozens of books. Recent topics as an author include the brands Gatorade and Nike. She lives in Minneapolis, Minnesota.